# Sustainable Agriculture and Globalization

# Sustainable Agriculture and Globalization

Dr. N.K. Rajan

AMIGA PRESS. INC
Delhi -110051

**AMIGA PRESS. INC**

H.O :- 28-A, Indra Park, Chander Nagar

Delhi :- 110051 (India)

Tel :- 011-22504028, +91-98736-57000

Email :- amigapress@gmail.com

amigaeditorial@gmail.com

amigapub@gmail.com

Website :- www.amigapress@co.in

This edition is published by Amiga Press.Inc

ISBN : 978-93-84533-35-9



First Published 2018



PRINTED : THOMSON PRESS INDIA

# Preface

Agriculture in India is one of the most prominent sectors in its economy. Agriculture and allied sectors like forestry, logging and fishing accounted for 18.6% of the GDP in 2005 and employed 60% of the country's population. It accounts for 8.56% of India's exports. About 43% of India's geographical area is used for agricultural activity.

Despite a steady decline of its share in the GDP, agriculture is still the largest economic sector and plays a significant role in the overall socio-economic development of India. The monsoons play a critical role in the Indian sub-continent's agriculture in determining whether the harvest will be bountiful, average, or poor in any given year. The entire rainfall in the sub-continent is concentrated in the few monsoon months.

Agricultural development has three distinct but related dimensions: the physical-technical, the economic-financial, and the institution-human. The physical-technical dimension addresses land utilization, agricultural technologies, research and extension, agricultural inputs, farm-to-market access, productivity and production maximization, and so on. Agricultural development from the economic-financial perspective is concerned with costs, factors of production, terms if trade, pricing policies, subsidies, incentives, credit, return on investment, market mechanisms, and the like.

The institutional-human dimension looks at knowledge and skills, organization and management, training, implementation capacity, social relations, politics, communication, motivation, participation, local government, public-private sector linkages, culture and values, historical experience, and so on. Agricultural development, as with much of the development field, has tended to be the province of the technical and economic specialists. Their language and concerns have dominated the analysis of rural and agricultural sector issues, the specification of agricultural development objectives, and the design of programmes and projects to achieve those objectives. The institutional-human dimension has frequently been either completely ignored or treated as a source of problems or constraints to achieving technical and economic targets, a residual category for anything not defined as technical or economic.

As with 'sustainable development' there is no precise, universally accepted definition of 'food security'. In broad principle a state of food security exists where all individuals, families and communities enjoy consistent access to foods that in quantity, quality and biochemical composition provide hygienic, nutritional adequacy. The Brundtland Panel states: "Food security requires secure ownership of, or access to, food resources and income earning activities, including reserves and assets to offset risks, to ease shocks and meet all contingencies". This may be interpreted as all people having access to stocks and flows of food and cash sufficient to satisfy their basic nutritional needs. Everyone in the world need not be a subsistence farmer, but those who do not produce foods in quantities sufficient to satisfy their families' needs, must possess the means to acquire foods nutritionally adequate, readily accessible and affordable.

The globalization of agriculture has resulted in less food being produced around the world, less diverse food being grown, less food reaching the poor and hungry and less food security (in the sense of access to safe and nutritious food) even in developed nations.

Genetically modified food finds its way into groceries in the United States without notice to the consumers. Beef tainted with BSE, meats and vegetables contaminated with e. coli, water contaminated with phosphates, nitrates, pesticides, fungicides, herbicides, hormones, the list goes on and on.

The gobalization of agriculture represents the very antithesis of sustainability. It is driven by short sighted greed based on corporate profits and the exploitation of both natural and human resources, and it rests on a myopic need to preserve culture, as it has been defined in the past half a century without thought to our biological survival.

This book provides deep insight to various dimensions of issues relating to the subject.

—*Editor*

# Contents

# 1

# Rural Development in Agriculture

## AGRICULTURE IN INDIA

Agriculture in India is one of the most prominent sectors in its economy. Agriculture and allied sectors like forestry, logging and fishing accounted for 18.6% of the GDP in 2005 and employed 60% of the country's population. It accounts for 8.56% of India's exports. About 43% of India's geographical area is used for agricultural activity.

Despite a steady decline of its share in the GDP, agriculture is still the largest economic sector and plays a significant role in the overall socio-economic development of India. The monsoons play a critical role in the Indian sub-continent's agriculture in determining whether the harvest will be bountiful, average, or poor in any given year. The entire rainfall in the sub-continent is concentrated in the few monsoon months.

Agriculture in India is constitutionally the responsibility of the states rather than the central government. The central government's role is in formulating policy and providing financial resources for agriculture.

The government administers prices of essential commodities to protect farmer's interests. It also administers other commodities which are produced by government-controlled companies, like petroleum, coal, nitrogenous fertilizers, etc. Other than these, most agricultural commodity markets operate under the normal forces of demand and supply.

## HISTORY OF INDIAN AGRICULTURE

Evidence of the presence of wheat and some legumes in the 6th millennium BC have been found in the Indus Valley. Oranges were cultivated in the same millennium. The crops grown in the valley around 4000 BC were typically wheat, peas, sesame seed, barley, dates and mangoes. By 3500 BC cotton growing and cotton textiles were quite advanced in the valley. By 3000 BC farming of rice had started. Other monsoon crops of importance of the time was cane sugar. By 2500 BC, rice was an important component of the staple diet in Mohenjodaro near the Arabian Sea. The Indus Plain had rich alluvial deposits which came down the Indus River in annual floods. This helped sustain farming that formed basis of the Indus Valley Civilization at Harappa. The people built dams and drainage systems for the crops. By 2000 BC tea, bananas and apples were being cultivated in India. There was coconut trade with East Africa in 200 AD. By 500 AD, egg plants were being cultivated.

## Agricultural Movements

### *Green Revolution*

After independence, considering India's growing population, the government took steps to increase the food production. Yields per unit area of all crops have grown since 1950. The 1970s saw a huge increase in India's wheat production. This is known as the Indian Green Revolution. Reasons for the growth are the special emphasis placed on agriculture and steady improvements in irrigation, technology, application of modern agricultural practices and provision of agricultural credit and subsidies

### *Operation Flood*

Operation Flood was the name of a rural development programme started by the National Dairy Development Board (NDDB) in 1970 with the objective of creating a nationwide milk grid. This movement followed the Indian green revolution and helped in alleviating poverty and famine levels from dangerous proportions in India during the era. It resulted in India becoming the largest producer of milk and milk products, so it is also called the White Revolution of India.

## Production

India is ranked second in the world in terms of agricultural output. India is the largest producer in the world of milk, cashew nuts, coconuts, tea, ginger, turmeric and black pepper. It also has the world's largest cattle population (193 million). India ranks second worldwide in farm output. It is the second largest producer of wheat, rice, sugar, groundnut and inland fish. It is the third largest producer of tobacco. India accounts for 10 per cent of the world fruit production with first rank in the production of banana and sapota. Given below is a chart of trend of output of cereals and major foodgrains as published by the Department of Food and Public Distribution with figures in tonnes.

## AGRICULTURAL GROWTH AND PERFORMANCE

While the above analysis has provided a general view of the impact of economic reforms, this chaper examines agricultural growth and performance in the states of Bihar, Karnataka, Tamil Nadu and Punjab with their attendant policy implications.

The yields for various crops in these states differ greatly. While Tamil Nadu had the highest yield in rice, oil seeds and sugarcane, Punjab enjoyed the highest yields in wheat, coarse cereals, pulses and food grains. Karnataka on the other hand is seen to do well in cotton and Bihar performed quite well in pulses and coarse cereals. Further analysis and findings by Kalirajan and others (2001) show that Punjab had made remarkable achievements on the agricultural front while Bihar had remained stagnant in the last two decades, with Karnataka and Tamil Nadu showing moderate achievement. Clearly, differences in physical endowments, climatic conditions and institutional characteristics are some of the reasons for the varying productivity performance. Thus, having across the board economic reforms is likely to work less effectively than state-specific policy measures that enable each state's agricultural yields to reach their full potential. The comparative advantage of each state's agricultural production should be determined and with inter-state restrictions removed, total agricultural output would see a very significant increase.

For example, Karnataka with less favourable soil and water resources should be given incentives to concentrate on agro-processed products and corporate agriculture in horticulture, floriculture and animal husbandry, or to undertake watershed development to help with dry land agriculture. Many studies have indicated that with watershed areas, productivity growth has been mainly due to seed and fertilizer use. Thus, this state has to be given input subsidies for high yielding seed varieties but at the same time, the farmers need to be educated on the over use of chemical fertilizers.

With Bihar, agricultural performance is problematic on many fronts. First, although demographic pressure has increased and agricultural technology has improved, most of the uncultivated land is concentrated in southern Bihar, where irrigation facilities have not kept pace and the soil is of poor quality. Given the physiography of southern Bihar, wells are also unsuitable and thus the dominant mode of irrigation has been through tanks whose expansion and maintenance has been neglected. Second, the infrastructural facilities of Bihar have been lagging as seen by the infrastructure development. Due to infrastructural bottlenecks, availability of modern goods and services has not increased or their supply remains costly or unreliable.

Third, agriculture in Bihar is dominated by small and marginal farmers and the prevalence of mass poverty is largely related to the backwardness of agriculture. Fourth and importantly, the state agricultural policies in Bihar are in dire need of review. The semi-feudal production condition still exists in rural areas and the ineffective protection of tenancy rights has hindered agricultural growth. The slow pace of land consolidation reflects inadequate financial outlays and a shortage of manpower. Kalirajan and others (2001) note that marketing and extension services in Bihar are also rather weak compared to the other states.

Punjab on the other hand, was one of the few states which enjoyed the success of land reforms and the high priority of investment in rural infrastructure. Also, the irrigation base of the small and medium sized farms was comparable to that of large farms. In addition, the Punjab Agricultural University at Ludhiana contributed to the development of new seed varieties.

However, there are clear signs of a decline in crop yields since the 1990s and this has been associated with the increasing use of fertilizers and excessive water use which have increased the unit cost of production as a result of declining soil quality. Hence, care is needed when providing further input subsidies in fertilizer and water use. Another related fact is the steep increase in wages in Punjab and in the absence of productivity increases, the cost increase has affected the profitability of farmers. With Tamil Nadu, the main crop has been rice as this state is blessed with two monsoons. But from 1992-1997, there has been a steady decline in the areas irrigated by canals and an increase in well-irrigated areas while the use of tanks remains an unreliable source of irrigation. However, major improvements in about 10 rice varieties released in the early 1990s can be expected to improve productivity growth in rice production although pests and diseases as well as imbalance in the use of fertilizers are major constraints. Thus Tamil Nadu could do with subsidies of pesticides and farmers should be educated on the more effective use of fertilizers to obtain high yields. Interestingly, the cropping pattern of late has shown increasing substitution of food crops by commercial crops but there is concern that the benefits will reach farmers only with the development of adequate infrastructure such as roads and markets. However, that Tamil Nadu has a higher index than the all India average of infrastructure.

## *Challenges of Globalization*

It is important to realize that globalization poses many challenges to a developing country like India, which had relied on a state directed and regulated policy regime for more than four decades. In moving to a more open, market-based economy there are many transitional problems that the country has to manage. The Government must play a pro-active role in facilitating the globalization process so that the opportunity sets for the economic agents are widened and the adverse effects of globalization are minimized. The Indian Government must also prepare the necessary information base and develop its capacity to articulate India's concerns and policy trade-offs in the international forums for multilateral trade and environmental negotiations.

In addition, the Government should embark on an extensive programme to educate farmers on the need to meet the standards required in the export markets. In fact, India needs to seek technical assistance in creating the capacity for meeting such standards and to consider watershed developments for environmental considerations. Equally important is the need to disseminate information about possible export markets to farmers, so that market access is achieved at minimum cost. Given the requisite information about markets and profitability, the likelihood of farmers investing in post-harvest and processing technologies and storage and efficient transportation arrangements as well as developing supporting infrastructure is very high.

Although the brave and bold move by India to reduce the tariff rate for agricultural products from 113 per cent in 1990-1991 to 26 per cent in 1997-1998 deserves to be applauded, the question of whether India is ready to compete in world markets remains to be seen. The infant industry argument may still hold for India to shield itself from external competition but one can easily question the length of time that is required to that end. Also, a delay in opening up to foreign trade has the danger that local producers may become too complacent and never be ready for competition.

As India opens up externally, it is also expected to face vulnerability in the wider international price fluctuations and thus Acharya (1998) claims that a minimum price support scheme is important. These prices can also act as a signal to adopt modern inputs and invest in yield-raising infrastructure for increasing production. For instance, keeping basic staple food grains at reasonable prices would induce farmers to switch over to high value crops. However, during the 1990s, Kalirajan and others (2001) shows that procurement prices especially for rice and wheat have been increasing faster than the general price level. Such high prices along with guaranteed purchases by the Food Corporation of India have pushed up market prices. These higher prices are partly responsible for the large buffer stocks with the Food Corporation. If this trend continues, India's comparative advantage will be eroded.

With openness and high price instability, unstable export revenue can also be expected. One way of reducing such risk is for India to diversify her agricultural exports. For example, since 1990, even in commodities such as tea, coffee, cocoa and spices, where India is supposed to have a comparative advantage international prices have been unstable. Besides increasing the type of exports to obtain more export revenue, India should also seriously consider exporting more value added agricultural products through agro-processing such as processed vegetables, fruits, fish and meat products given that export or even local demand for basic agricultural products would decline as incomes rise. The move to higher value added activities within the agricultural sector also spells greater opportunities for industrialization and vice versa as borne by Kalirajan and Shand's (1997) findings of a bi-directional relationship between agriculture and industry for most Indian states. On the other hand, Sivakumar and others (1999) establish empirical evidence of high forward linkages of agriculture due to the presence of agro-industries while Satyasai and Viswanathan (1999) show the significance of the spillover effects to the industrial sector via the intensive use of purchased inputs in the agricultural sector.

The lack or slow pace of internal or domestic liberalization is also seen to hinder the possible gains from external or trade liberalization. For example, although central zoning restrictions have been abolished, state government restrictions on inter-state and even inter-district restrictions on marketing and movement of goods still exist in many cases. This interferes with the benefits from crop specialization and economies of scale arising from comparative advantage. The land market is another example of distortion whereby land ceilings exist preventing the operation of large-sized farms. This has led to the emergence of a large number of small economically unviable land holdings. The easy leasing of land should be permitted with assurance of resumption. Yet another problem lies with the insufficiency of credit to agriculture. From 1995-1996, the Rural Infrastructure Development Fund was set up to allocate funds for the completion of projects and the government has committed itself to strengthening the cooperative credit structure through substantial refinancing and restructuring of the Regional Rural Banks. However, as mentioned earlier, due to varying institutional factors in the Indian states, these domestic reforms can be expected to yield quite different results.

Although India missed the opportunity to open up two decades ago, its attempts to do so now must be regarded as better late than never. Others such as Desai (1999) observe that, "the logic of the global economy as well as India's interests dictate that India become proactive in its liberalization policies. India must liberalize not because it has no choice but because it is the best choice".

His lament that India has adopted a 'victim mentality' when it really needs to adopt a 'winner mentality' has become less of a concern as over time, India has shown commitment to stay on the bandwagon of globalization. Having realized that globalization is a necessary but not a sufficient condition for high growth production, India has undertaken economic reforms, both internal and external. However, it must be ensured that these reforms are synchronized so that the pace of both reforms is set right in order to work hand in hand to promote agricultural productivity growth.

Thus, training the farmers and educating them appropriately to change their mindset and reorienting them to take up new activities or adopt foreign technology is of utmost importance. In this context, it is necessary to involve non-governmental organizations in training and mobilizing the rural poor to face the challenge of liberalization.

Also, with domestic economic reforms, more care needs to be exercised to draw up state-specific liberalization measures to maximize their benefits. Lastly, in the implementation of these reforms for successful globalization, one crucial element, not entirely within control is the need for good governance and stability in the political and economic environment. Political leaders who are the ultimate decision makers in these matters need to examine their own role dispassionately.

It is quite apparent that at this relatively early stage, there is little observable evidence of gains to India's agricultural performance after opening up. However, there could easily be benefits that have not yet surfaced, or are yet to be identified and perhaps too difficult or intangible to measure.

Whatever the case, it is highly likely that it is too soon to assess the full impact of globalization and economic reforms. Furthermore, the process of liberalization has been gradual and remains incomplete. For example, the complete removal of quantitative restrictions after March 2001 will have provided an opportunity for Indian farmers to tap world markets and, if they are successful, results should start to become evident soon. Export promotion via the development of export and trading houses as well as effective liberalizing export promotion zone schemes for agriculture are fairly recent measures and only time will tell as to how effective these measures are. Other possibilities such as agro-industry parks for promoting exports are also in the pipeline.

## AGRICULTURE, GROWTH AND POVERTY REDUCTION

History shows that different rates of poverty reduction over the past 40 years have been closely related to differences in agricultural performance – particularly the rate of growth of agricultural productivity. In Asia, the rapid productivity gains of the Green Revolution increased producers' incomes, raised labourers' wages and lowered the price of food. In addition, new livelihood opportunities were generated when success in agriculture provided the basis for economic diversification. However, despite decades of investment in new agricultural technology and rural development, hunger and poverty continue to plague large areas of the developing world. The problem is particularly acute in sub-Saharan Africa, where progress towards the Millennium Development Goals is slowest.

While increasing agricultural productivity perhaps remains the single most important determinant of economic growth and poverty reduction, serious doubts are emerging as to whether agricultural productivity can be further increased, where it is most needed, and what part, if any, small-scale farming will play in the future. Development experts need greater understanding of the links between agricultural productivity and poverty. They also need to assess just how far they have changed and the extent to which small-scale agriculture can remain a ladder out of poverty for millions of poor people living in rural areas. Links between agriculture and poverty reduction are forged through four 'transmission mechanisms':

- direct and relatively immediate impact of improved agricultural performance on rural incomes;
- impact of cheaper food for both urban and rural poor;
- agriculture's contribution to growth and the generation of economic opportunity in the non-farm sector; and
- agriculture's fundamental role in stimulating and sustaining economic transition, as countries (and poor people's livelihoods) shift away from being primarily agricultural towards a broader base of manufacturing and services.

However, the potential for future poverty reduction through these transmission mechanisms depends on the extent to which agricultural productivity can be increased where it is most needed or, more simply and directly, the extent to which the Green Revolution can be replicated in Africa. Whether and how this might be achieved has been the subject of protracted debate amongst researchers and policy makers. The main questions in this debate are:

- Where to focus development efforts – high potential areas where development options are greater or poorer areas where the potential and options are less but the incidence of poverty is greatest?
- Who to focus on – the poorest smallholders who produce for subsistence and have limited engagement with markets, or larger-scale farmers

whose success can be an engine of growth and generate wealth and jobs for rural households?

- What to do in less favoured areas – should agriculture be promoted amongst poor households in areas of low agricultural potential that are remote from markets and inputs supplies? Or should the focus be on employment generation to enable poor households to buy in food?
- What role for technology – should primacy in development efforts be given to yield-raising technology or to less intensive approaches that minimise variation?
- Which crops – should the emphasis be on crops that will be largely consumed within the households, or on income-generating cash crops?

Contrasting answers to these questions produce different policy positions. Those who are optimistic about the role of smallholders emphasise the need for increased direct investment in agriculture and rural development, with support focusing on creating institutions that will encourage and support smallholder-led agricultural development. They believe it is important to improve the productivity of staple food crops that are not traded internationally, but consumed by the poor and traded locally.

In contrast, smallholder pessimists emphasise the need to achieve the best outcome possible from rapidly changing global markets for agricultural produce. They believe that commercial production (probably on larger farms) of non-staple cash crops should be encouraged, particularly those that result in robust links to the non-farm sector where the main source of employment for the rural poor would be found. Influencing international policy processes to ensure access to developed country markets is a priority for this position, as is improving the human capital assets of the rural poor so they can take up opportunities in growth areas as they occur.

## What is the Issue?

Poverty has fallen rapidly over the past 40 years, but at different rates around the world. Asia has achieved the most rapid poverty reduction, particularly China, but also India and South East Asia. In contrast, little if any progress has been made in sub-Saharan Africa, where the number of people living on less than one dollar a day – the internationally agreed definition of absolute poverty – has doubled over the past 20 years (World Bank, 2004a).

Historically, rates of poverty reduction have been very closely related to agricultural performance – particularly to the rate of growth of agricultural productivity. In simple terms, this indicates that the countries that have increased their agricultural productivity the most have also achieved the greatest reductions in poverty. In Asia rapid productivity gains, achieved largely through the technological advances of the Green Revolution, provided a 'fast-track' route out of poverty by directly increasing producers' incomes and labourers' wages, by lowering the price of food and by generating new livelihood opportunities as success in agriculture provided the basis for economic diversification.

However, despite decades of investment in new agricultural technology and rural development, hunger and poverty continue to plague large areas of the developing world.

The problem is particularly acute where people depend on rain-fed agriculture, in particular sub-Saharan Africa, where the impact of new technologies has been less apparent and agricultural productivity has generally stagnated and even fallen in some areas.

Achieving the Millennium Development Goal (MDG) of halving the proportion of people living in absolute poverty by 2015 will depend largely on increasing agricultural productivity, which remains perhaps the single most important determinant of economic growth and poverty reduction. This fact is not lost on developing countries or their development agency partners, who are seeking ways to stimulate agricultural development. But serious doubts are emerging as to whether agricultural productivity can be increased where it is needed most, and what part, if any, small-scale farming will play in the future. Underlying such doubts is a concern that the context in which small-scale agriculture could achieve productivity gains today is very different to that which prevailed in Asia during the halcyon days of the Green Revolution. Inherent differences in production capabilities and quite fundamental changes in the international agricultural context are combining to create a set of circumstances that are probably far less conducive to achieving the type of productivity transformation witnessed in Asia.

Development experts need greater understanding of the links between agricultural productivity and poverty. They also need to assess just how far they have changed and the extent to which small-scale agriculture can remain a ladder out of poverty for millions of poor people living in rural areas.

## A SAGA OF SUCCESS

From a nation dependent on food imports to feed its population, India today is not only self-sufficient in grain production but also has a substantial reserve. The progress made by agriculture in the last four decades has been one of the biggest success stories of free India. Agriculture and allied activities constitute the single largest contributor to the Gross Domestic Product, almost 33 percent of it. Agriculture is the means of livelihood of about two-thirds of the workforce in the country.

This increase in agricultural production has been brought about by bringing additional area under cultivation, extension of irrigation facilities, the use of improved high-yielding variety of seeds, better techniques evolved through agricultural research, water management, and plant protection through judicious use of fertilisers, pesticides and cropping practices.

## Crops

The 1970s saw a multi-fold increase in wheat production that heralded the Green Revolution. In the next decade rice production rose significantly; in 1995-96, rice production was 79.6 million tonnes. The total grain production crossed 211 million tonnes in 2001-02, a big leap from 51 million tonnes in 1950-51. To carry improved technologies to farmers, a National Pulse Development Programme, covering 13 states, was launched in 1986. Efforts to boost pulse production were augmented further by the Special Food Production Programme. In 2001-02, pulse production was 13.52 million tonnes. With some states offering more than the statutory minimum price, sugar cane production also received a boost, and in 2001-02 a record 292.2 million tonnes was registered.

## Irrigation

As efforts continued to increase the irrigation potential in the country, the last 40 years saw the gross irrigated area reach 85 million hectares. Flood forecasting has become an important activity over the years. Over 500 hydrological stations collect and transmit data through 400 wireless stations for issuing forecasts for 157 sites. About 5,000 forecasts are issued in a year with 94 percent accuracy. The country also receives international support, with the World Bank as a primary source, for developing its water resources. International cooperation is also envisaged in setting up a National Centre for Information on Water and Power. As there is a broad seismic belt in the country, particularly along the Himalayan, and the Kutch region and parts of Maharashtra, a scheme is being evolved to collect all data on seismic activity at various dam sites.

## Fertilisers

The fertiliser industry in India has grown tremendously in the last 30 years. The Government is keen to see that fertiliser reaches the farmers in the remote and hilly areas. It has been decided to decontrol the prices, distribution and movement of phosphatic and potassic fertilisers. Steps have been taken to ensure an increase in the supply of non-chemical fertilisers at reasonable prices. There are 66 fertiliser quality control laboratories in the country. Since biofertilisers are regarded as an effective, cheap and renewable supplement to chemical fertilisers, the Government is implementing a National Project on Development and Use of Bio-fertilisers. Under this scheme, one national and six regional centres for organising training, demonstrating programmes and quality testing of bio-fertilisers has been taken up. It was a challenging decision of the Government to take Bombay High gas through a 1,700 km pipeline to feed fertiliser plants located in the consumption centres of North India.

However, the major policy which has ensured the growth of the fertiliser industry is the thrust on accelerating fertiliser consumption by fixing, on the one hand, low and uniform price for fertilisers, and on the other hand providing the manufacturers adequate compensation through the retention price and subsidy scheme. As expected, fertiliser nutrient demand has gone up from 0.29 million tonnes in 1960-61 to 16.7 million tonnes at the end of 2000-01, compared to 12.15 million tonnes during 1992-93.

### Fisheries

Fish production achieved an all-time high of 5.6 million tonnes at the end of 2001-02. Programmes that have helped boost production include the National Programme of Developing Fish Seeds, Fish Farmers' Development Agencies and Brackish Water Fish Farmers' Development Agencies. The Central Institute of Fisheries Nautical and Engineering Training trains the necessary manpower. To diversify fishing methods and introduce processed fish products on a semi-commercial scale, an Integrated Fisheries Project has been launched. A National Fisheries Advisory Board has also been established.

## FOOD PROCESSING

The Ministry of Food Processing Industries, set up in July 1988, is the central agency of the Government responsible for developing a strong and vibrant food-processing sector with a view to create increased job opportunities in rural areas, enable the farmers to reap benefit from modern technology, create surplus for exports and stimulating demand for processed food. A new seeds policy has been adopted to provide access to high-quality seeds and plant material for vegetables, fruit, flowers, oilseeds and pulses, without in any way compromising quarantine conditions. Initiatives have been taken to encourage private sector investment in the food processing industry.

## AGRICULTURAL RESEARCH

The apex body for education, research, training and transfer of technology in the field of agriculture is the Indian Council of Agricultural Research (ICAR), established in 1929. India's transformation from a food deficit to a food surplus country is largely due to ICAR's smooth and rapid transfer of farm technology from the laboratory to the land.

ICAR discharges its responsibilities through 43 research institutes, four national research bureaux, 20 national research centres, nine project directorates, 70 all-India coordinated research projects, and 109 Krishi Vigyan Kendras (farm science centres). Besides, the programme of Agricultural Education is coordinated by ICAR with the curricula and other normative guidance given to the 26 agricultural universities and four national research institutes.

## OILSEEDS PRODUCTION

A Technology Mission on Oilseeds was launched in 1986 to increase production of oilseeds in the country and attain self-sufficiency. Pulses were brought under the Technology Mission in 1990. After the setting up of the Technology Mission, there has been consistent improvement in the production of oilseeds. The oilseed production, which was 108.3 lakh tonnes in 1985-86, has increased to 247.30 lakh tonnes in 1998-99, the highest production so far. However, due to unfavourable weather conditions prevailing in the major oilseeds-growing states, the oilseed production during 1999-2000 and 2000-01 suffered some decline.

The increase in production was largely contributed by soyabean, rapeseed and mustard. Production of pulses has seen many ups and downs, which is expected to be checked under the mission. The country grows mainly nine oilseeds, with groundnut, rapeseed and mustard accounting for 62 percent of total production. Lately, soyabean and sunflower have shown major growth potential.

## DRINKING WATER

A Technology Mission on Drinking Water and Related Water Management has been constituted to cover the residual problem villages and provide potable water at 40 litres per capita per day, and 70 litres per capita per day in desert areas inclusive of 30 litres for cattle. The mission is tackling the problem through 55 mini-missions in project districts and countrywide problem-oriented sub-missions. A Village Level Operation and Maintenance (VLOM) pump called India Mark-II has been developed and is being exported to 40 countries.

## INSTABILITY ASSOCIATED WITH IMPROVED AGRICULTURAL TECHNOLOGY

There has been much concern over the possibility that improved crop varieties are associated with increased output instability in addition to higher yields. If this is the case, it could deter adoption of more productive technology and hence retard agricultural development.

### Instability at the Farm Level

In discussions of farmers' adoption of HYVs, it is often assumed that new seeds have more volatile yields. Indigenous landraces often have the characteristic of being fairly unresponsive to improved management and increased inputs, but also robust in the face of unfavorable agroclimatic conditions, such as drought. HYVs, on the other hand, often are characterized as being highly responsive to management and inputs, but very low yielding in the event of bad weather. If this is the case, then plant breeders must breed on the basis of one strategy (high average yield) for high potential areas and another (risk minimization) for low potential areas.

Research on pearl millet, which is commonly grown in unfavorable agroclimatic areas, suggests that the above characterization of traditional varieties and HYVs is not necessarily correct. Witcombe (1989), in a study of Pakistan and India, found that good performance of a particular millet seed over all environments appears to indicate good performance in environments of low potential. In three years out of his four year study, the highest yielding entry across all environments was also one of the two highest yielding entries in the lowest yielding environment. These results suggest that the typical plant breeder's strategy of selecting among the highest yielding seeds across all environments is satisfactory. Farmers who adopt HYVs do not necessarily subject themselves to greater risk of catastrophic loss in the event of bad weather.

Farmers' Strategies to Reduce and Cope with Risk. Walker and Jodha (1986) point out that dryland farmers have various methods to reduce their exposure to crop production risk. Cultural practices play an important risk-reducing role; they include planting different crops with relatively low covariate yield (either in an intercrop or on separate fields); diversifying spatially by operating multiple plots with different environmental characteristics; and staggering planting dates in the face of variable rainfall patterns. Sharecropping is a common tenancy arrangement that distributes risk between the tenant and landlord. Many farmers have multiple sources of income, reducing risk if they have low covariation.

Farmers also have various mechanisms to cope with risk that they cannot eliminate. For example, they can borrow from local stores or money lenders, draw down food stocks or savings, sell assets, obtain transfers from relatives, participate in government relief programs, or migrate. Most of these options are not particularly desirable; for especially poor people they can be quite devastating: selling assets or going into debt may make a family permanently worse off even after drought is over. Jodha (198_) indicates that some families will reduce their food consumption as much as possible before parting with their assets; this has obvious negative short term health implications that are particularly severe for those who consume only minimum requirements to begin with.

Effect of Risk on Technology Adoption. While weather-related risk undoubtedly presents great hardship for a very large number of people, Walker (1989b) indicated that it may not be as important as generally believed in adoption of new technologies in rainfed areas.

The tradeoff between expected income and the variance of that income suggests that given farmers' measured risk preferences, the overall effect on adoption decisions is modest. Also, Walker stressed that yield instability does not translate into major variability in income due to farmers' mechanisms to absorb risk. As mentioned above, multiple sources of income and diverse cropping patterns mean that yield variability of one crop only affects a portion of the income from crops. Equally important, in many cases farmers can adjust the area under each crop depending on the weather at the start of the season. For example, dryland rabi (postrainy season) sorghum farmers in black soil areas know at the start of the season how much moisture is available and adjust their cropped area accordingly. Similarly, castor farmers in Andhra Pradesh know that pest attacks are more prevalent when the rainy season begins late, so they plant less castor. And farmers with irrigation from tanks or wells know roughly how much water will be available in the postrainy season, so they adjust planted area accordingly. Hazell and Ramasamy (1991), for example, found that sharply reduced paddy production in North Arcot, Tamil Nadu, in the drought year 1982-83 resulted from a fall in area, not yield. Not surprisingly, Walker (1989b) also found that variations in cultivated area This is why SWC programs are sometimes referred to as "drought-proofing" measures.

exceeded those in yield. He also found that given farmers' diversified farming and livelihood strategies, the large variations in yield translated into only small variations in income. The variations were so small that even if plant breeders could develop varieties with perfect stability (zero variation), the contribution to household income stability would amount to less than 1% for most of the crops studied, and a maximum of 2.9% for paddy (which, paradoxically, had the least yield variability to begin with).

Walker's (1989b) results demonstrate the importance of looking beyond yield variability of a single crop to variations in all the crops in a given household, village or region, and beyond crop variation to income variation. From this perspective, yield instability does not appear to be a major determinant of adoption, and hence not a top priority for plant breeders. He pointed out (in his 1989a study) that policies related to international trade and storage between surplus and deficit years can be more cost-effective in coping with increasing yield instability.

Soil and water conservation (SWC) investments also are associated with a variety of risks. First, erosion itself is a matter of risk. For example, some plots may be at risk of productivity loss from continuous, gradual erosion, but others may be more susceptible to significant erosion only in the event of a once-in-five-years or once-in-fifty-years storm. For such lands, in normal years there is no gain from investments to reduce erosion, but in exceptional years the gain — actually the avoided loss — may be quite high. Second, often SWC practices serve to conserve soil moisture as well as reduce erosion. Soil moisture retention is more likely to offer immediate, productivity-increasing benefits than erosion prevention. However, these benefits may exceed the costs only when rainfall is unusually low or unevenly distributed, so that moisture stress constrains productivity.12 In a good rainfall year, on the other hand, short term gross returns to increased moisture retention may be low or zero, and in a very high rainfall year they may even be negative if they lead to waterlogging. Little information is available regarding the impact of risk on Indian farmers' SWC investments.

## Instability at the Aggregate Level

Ramakrishna (1993) used a log function to calculate CVs to compare instability of production, yield and area cultivated for cereals between the pre-green revolution and green revolution periods. Ramakrishna found that for food grains, output was more stable during the green revolution than before. Yields were about equally stable between the two periods, and area became more stable. For other crops, on the other hand, Ramakrishna found that output and yield instability both increased during the green revolution, while area again became more stable. Combining food grains and other crops, output and yield instability increased slightly during the green revolution, while area instability was constant. On the whole, Ramakrishna's data suggest that the green revolution had little impact on agricultural instability.

Hanumantha Rao (1994) took a slightly different approach and found different results. Instead of taking CVs of output levels, he took the standard deviation of annual output growth rates as a measure of instability. He found that the standard deviation of output growth of food grains was 8.1 in 1950-51 but 11.4 between 1968-85, and took this to indicate that output instability increased with the green revolution. He also found that the standard deviation rose from 9.4 in the first decade of the green revolution to 11.8 in the second decade. Hanumantha Rao did not provide detailed figures regarding yield and area instability. His figures do not really offer a good comparison to those of Ramakrishna, because his pre-green revolution figures are based only on a single year. Also, it is difficult to compare the magnitude of Hanumantha Rao's figures (expressed as the standard deviation of output growth rates) to Ramakrishna's (expressed as the CV of the level of output). Since we do not know the growth rates from which Hanumantha Rao calculated the standard deviation, we cannot relate his standard deviations to Ramakrishna's CVs. Hazell (1982) examined cereal crop production between 1954-55 and 1964-65 (before the green revolution) and 1967-68 and 1977-78 (during the green revolution, after the introduction of high yielding varieties).13 He found that the CV of production increased by about 50%, from 0.04 to 0.059 between the two periods. Hazell hypothesized that if the increased instability were due to HYVs, variances in production within states would have to rise. However, he found that changes in yield covariances were much more important than changes in yield variances. Only about 18 percent of the increase in variance of total cereal production resulted from changes in crop production variances; the remaining 82 percent was explained by changes in covariances; interstate covariances within crops contributed 41 percent to the change in variance in total cereal production. As a result, Hazell concluded that HYVs were probably not the primary cause of increased variability. In a later paper, Hazell (1984) suggested that HYVs could possibly affect yield covariances of maize in India and the United States, because the narrower genetic base of improved varieties would make them susceptible to common yield inhibitors such as pests and diseases. This would then be a contributing factor to greater variability in national ceral production.

Walker (1989a) studied changes in yield and output variability of sorghum and pearl millet in India resulting from the spread of HYVs. This study followed Hazell's 1984 study relating the rising covariance of yields between states to the spread of HYVs; it used districtlevel data for the 48 largest sorghum districts and the 40 largest pearl millet districts. Walker found that variability had indeed increased significantly between the period 1956-57 to 1967- 68 and 1968-69 to

1979-80, from 8% to 16% for sorghum and 11% to 34% for pearl millet. He also found that covariance of yields across districts was by far the most important factor in overall output variance of sorghum and pearl millet. Walker examined several possible causes of increased covariance, including 1) changes in rainfall covariance, 2) changes in irrigated area, and 3) adoption of HYVs. He found evidence that provided weak support for each of these possible sources of covariance. However, he also stressed that the contribution of HYVs to increased instability is dwarfed by their contribution to productivity, so he did not recommend changes in existing breeding strategies that develop HYVs for adoption over a large area.

Walker's (1989a) study helps clarify that numerous factors can contribute to changes in stability. Researchers who find that stability increased or decreased with the green revolution should hesitate before proclaiming that HYVs were the cause. Changes in cropping patterns, fluctuations in weather, and the quality of land on which a particular crop tends to be planted all can have an impact on the variations in yield. More importantly, variations in area may either counteract or reinforce those in yield; it is important to know the composition of output fluctuations between yield and area variations. The studies by Hazell and Walker also highlight the need to distinguish between variability at the regional or national levels from farm or field level risks. Because of the dominance of covariance relations in aggregate production data, variability can increase at the aggregate level even while farm level variability changes little or not at all. Although diets were stable, health problems increased due to a shortage of clean drinking water in the village resulting from the drought.

## DROUGHT RISK

The preceding discussion on output instability has underemphasized the risks associated with the specific problem of severe, prolonged drought. When severe drought occurs over a wide area, yield and income risks can be particularly great. In Hazell and Ramasamy's (1991) study of North Arcot district, Tamil Nadu, for example, they found that the 50% decline in paddy production in their study villages was replicated throughout the district and beyond. Incomes fell by 50% on average as even nonfarm incomes fell sharply. Consumption expenditure fell by 50% on average, and diet quality deteriorated. Clearly, the effects of drought were covariate across villages and sectors. This problem of covariate risk is particularly challenging because poor performance of one income source, or one location, cannot necessarily be compensated by better performance in another. As a result, people may face severe hardships in the event of drought.

In the Bidinger et al (1990) study, on the other hand, despite a similar 50% decline in income, dietary intake did not change compared to the pre-drought situation.14 People maintained their dietary intake through increases in temporary migration and consumption credit, and government rice subsidies that kept prices low and stable, enabling people to translate meager incomes into normal diets. The number of people who migrated only increased by about 10%, but they stayed away much longer on average. Interest rates did not show the expected increase as demand for consumption credit rose, even though the bulk of borrowing was done through the informal village credit market rather than the formal banking system, which contributed little. Distress sales of land and livestock were rare, but two families did sell all their assets and move permanently to become urban laborers.

## GOVERNMENT INTERVENTIONS TO MANAGE RISK

Severe droughts impact negatively on most rural households simultaneously and are therefore difficult to manage through traditional risk sharing and coping strategies. As a result, government policy may sometimes play an important role in helping farmers manage risk. While some government interventions are well-established, other ideas remain relatively untested. Likewise, even where farmers and rural communities have developed effective mechanisms to manage production risk, it is not clear what the costs of these mechanisms are in terms of reduced production efficiency. Perhaps market-based or government-sponsored alternatives can be introduced that protect farmers from the effects of risk but without requiring diverse income sources and fragmented agricultural holdings. Anderson and Hazell (1994) point out that more information is needed about the costs of these risk-reducing and coping mechanisms. We briefly review government measures to manage risk and discuss some additional possible government approaches.

Most of the existing drought management efforts. These are poverty allevation programs to cope with the effects of drought, namely employment programs and food subsidies. There is no need to repeat the discussion of these programs here. Bidinger et al (1990), stress the importance of food subsidies in mitigating the effects of drought in their study, but they also point out that a timely public works program could have prevented much of the unemployment and debt experienced by laborers in the village.

Crop insurance is provided by the public sector in many countries. The impetus for such programs often originates in governmental concern about catastrophic risks such as drought, or the desire to reduce the incidence of loan defaults to banks. With few exceptions, the financial performance of public crop insurers has been ruinous (Hazell 1992). To be financially viable without government subsideis, an insurer needs to keep the average value of its annual outgoings—indemnities plus administration costs—below the average value of the premiums it collects from farmers. In practice, many of the larger insurance programs pay out $2.50 or more for every dollar of premium they collect from farmers. The difference is paid by governments, at costs varying from $10 to $400 per insured hectare. Even at these levels of subsidy, many farmers are still reluctant to purchase insurance. As such, many crop insurance programs are compulsory, either for all farmers growing specified crops or for those who borrow from agricultural banks. The primary reason for the high cost of public crop insurance schemes is that they invariably attempt to insure risks that are prone to severe moral hazard problems, whereby farmers have incentive to lie about their production levels, or to allow their crops to fail in order to receive insurance payments (Hazell 1995a). These risks include many climate, disease and pest risks that are difficult to quantify and assess, and whose damage can be influenced by armers' management practices. The problem is aggravated by a common practice of insuring "target" yields rather than compensating for actual losses. But this is not the only reason for failure.

Another overwhelming factor is the incentive problem that arises oce the government establishes a pattern of guaranteeing the financial viability of an insurer. If the insurance staff know that any losses will automatically be covered by governmetn, they have little incentive to pursue sound insurance practices when setting premiums and assessing losses. In fact, they may find it profitable to collude with farmers in filing exaggerated or falsified claims. Yet another common reason for failure has been that governments undermine public insurers for political reasons. In Mexico, the total indemnities paid has borne a strong statistical relationship with the electoral cycle, increasing sharply immediately before and during election years, and falling off thereafter. In the USA, the government has repeately undermined the national crop insurer (FCIC) by providing direct assistance to producers in disaster areas. Why should farmers purchase crop insurance against major calamities (including drought) if they know that farm lobbies can usually apply the necessary political pressure to obtain direct assistance for them in times of need at no financial cost? Another reason for their high cost is that crop insurers tend to be too specialized, focusing on specific crops, regions and types of farmers, particularly when the insurance is tied to credit programs designed to serve particular target groups identified by the government. Without a well-diversified insurance portfolio, crop insurers are susceptible to covariability problems, a face the prospe t of sizable losss in some years. Since public insurers are rarely able to obtain commercial reinsurance or contingent loan arrangements, thsi specialization increasestheir dependence on the government. Crop insurances is provided by the public sector in many countries. The impetus for such programs often originates in governmental concern about catastrophic risks such as drought, or the desire to reduce the incidence of loan defaults to banks. With few exceptions, the financial performance of public crop insurers has been ruinous (Hazell 1992). To be financially viable without government subsidies, an insurer needs to keep the average value of its annual outgoings – indemnities plus administration costs – below the average value of the premiums it collects from farmers. In practice, many of the larger insurance programs pay out $2.50 or more for every dollar of premium they collect from farmers. The difference is paid by governments, at costs varying from $10 to $400 per insured hectare. Even at these levels of subsidy, many farmers are still reluctant to purchase insurance. As such, many crop-insurance programs are compulsory, either for all farmers growing specified crops, or for those who borrow from agricultural banks. The primary reason for the high cost of public crop-insurance schemes is that they invariably attempt to insure risks that are prone to severe moral hazard problems whereby farmers have incentive to lie about their production levels, or to allow their crops to fail in order to receive insurance payments (Hazell 1995a). These risks include many climate, disease and pest risks that are difficult to quantify and assess, and whose damage can be influenced by farmers' management practices. The problem is aggravated by a common practice of insuring "target" yields rather than compensating for actual losses. But this is not the only reason for failure.

Another overwhelming factor is the incentive problem that arises once the government establishes a pattern of guaranteeing the financial viability of an insurer. If the insurance staff know that any losses will automatically be covered by government, they have little incentive to pursue sound insurance practices when setting premiums and assessing losses. In fact, they may find it profitable to collude with farmers in filing exaggerated or falsified claims.

Yet another common reason for failure has been that governments undermine public insurers for political reasons. In Mexico, the total indemnities paid has borne a strong statistical relationship with the electoral cycle, increasing sharply immediately before and during election years, and falling off thereafter. In the USA, the government has repeatedly undermined the national crop insurer (FCIC) by providing direct assistance to producers in disaster areas. Why should farmers purchase crop insurance against major calamities (including drought) if they know that farm lobbies can usually apply the necessary political pressure to obtain direct assistance for them in times of need at no financial cost? Another reason for their high cost is that crop insurers tend to be too specialized, focusing on specific crops, regions and types of farmers, particularly when the insurance is tied to credit programs designed to serve particular target groups identified by the government. Without a well-diversified insurance portfolio, crop insurers are susceptible to covariability problems, and face the prospect of sizable losses in some years. Since public insurers are rarely able to obtain commercial reinsurance or contingent loan arrangements, this specialization increases their dependence on the government. Public crop insurers also tend to have high administration costs. This is partly because they often insure small-scale farmers, but also because crop-insurance work is very seasonal, and the absence of a well-diversified portfolio means that staff and field equipment are underemployed for significant parts of the year.

There is no convincing evidence that public subsidization of crop insurance has been socially beneficial. Indeed, social benefit-cost analyses of the Mexican and Japanese schemes show negligible social returns in relation to their high costs (Bassoco et al. 1986, Tsujii 1986). Nor is there much evidence that it has increased agricultural lending or benefited agricultural banks. In a rare study, Pomareda (1984) compared the performance of insured and uninsured loans in the portfolio of the Agricultural Development Bank (BDA) of Panama. Insured loans had slightly higher and more stable returns than uninsured loans. They were also repaid and cleared from the books closer to their expected duration. But the overall gains to the Bank were modest, and could have been achieved more easily at no cost to the government simply by allowing a 2 percent increase in the interest rate that BDA charged its borrowers. This would also have been cheaper for the borrowers than the premium rates they paid for the compulsory insurance.

In order to overcome the major problems associated with crop insurance, and to substantially reduce its administration costs, several authors have proposed area-based yield insurance (e.g. Halcrow, 1948; Dandekar, 1977; Miranda, 1991). Under this proposal, the crop yield for a *homogeneous region* is insured, and all insured farmers in the region pay the same premium and receive the same indemnity. Indemnities are paid whenever the average yield for the region falls below some critical level irrespective of the actual yields obtained by individual farmers. Premiums are calculated on the basis of year-to year variations in the average yield for the region, and would vary from one homogeneous region to another in accordance with differences in risk levels. This approach reduces moral hazard problems, and hence broadens the range of yield risks that can be viably insured. Moreover, since the premiums and indemnities are identical for all insured farmers in a region, it avoids the *adverse selection* problem. The latter refers to the situation in which farmers facing below-average risk tend to drop out of insurance programmes if they are charged premium rates based on average risk levels but are paid indemnities based on their own losses. Also, by eliminating the need for field inspections and loss assessments, the cost of administering an area-based scheme could be kept very low. Providing farmers pay their premium, it is not really necessary that they even grow the crop that they have insured.

Despite its appeal and its potential scope for reaching small-scale farmers, there are problems with the proposal. First, the insurance will be attractive to individual farmers only if their yields are highly correlated with the average yield for the region. Dandekar (1977) argues the homogenous areas can be defined in India in which inter-farm yield correlations are positive, but their is a growing body of micro-evidence showing that yield correlations between plots within the same village, or even within the same farm, are surprisingly low (e.g. Walker and Jodha, 1986). Small differences in ground contours, slope, and wind and sun exposure can lead to substantial differences in the yield damage caused by unfavorable climatic, pets and disease events, as can a few days difference in planting dates or the crop varieties grown.

Second, the scheme is subject to severe covariability problems. When the average yield is below the critical value in a region, all the insured farmers have to be compensated simultaneously. An unsubsidized insurer could hope to survive only if the scheme were to span a large number of regions with negative or positive but weakly correlated yields. The alternative of seeking commercial reinsurance seems unlikely given private insurers' reluctance to insure yields against a wide array of perils.

India introduced a national area-yield crop insurance scheme in 1985 (the Comprehensive Crop Insurance Scheme, CCIS), following a pilot phase from 1979 to 1984. The insurance is only available to farmers who borrow credit from financial institutions; and the indemnities are paid directly to the lending institution. Mishra (1994) provides evidence from Gujarat that the insurance did increase lending to small farmers, though repayment rates varied slightly. Unfortunately, the premium rates charged farmers are only a fraction of the rates needed to cover the indemnities paid, and even the small premiums charged (1-2% of coverage) are heavily subsidized by government. As a result, the financial performance of CCIS to date has been disastrous. The CCIS collected Rs 11,961 lakhs ($40 million) of premium during 1985-92, but paid Rs 90,163 lakhs ($303 million) indemnities over the same period (Joshi, nd). Considering that government subsidized about half the premium collected,

CCIS paid Rs 15 for each rupee of premium collected from farmers. The insurance program will clearly need to be reformed if it is to become a cost-effective risk management policy.

Another variant of area-based insurance is regional rainfall insurance. In this case the insurance pays out whenever the average rainfall for a region falls below some critical value. Rainfall is easier to measure than average yield, and with modern satellite imagery it is now possible to assess the soil moisture content for a region with the minimum of field inspection. This information could be used to confirm rainfall readings, or even used directly as the insured peril.

Unlike regional yield insurance, rainfall insurance is not tied to the performance of specific crops, Since most farm families rarely depend on a single crop for their total income, the attractiveness of rainfall insurance should depend more on how the total insurance is limited to the occurrence of severe drought or floods, it is likely that in semi-arid or floodprone areas the indemnities would coincide with catastrophic income outcomes for many rural families.

Drought insurance could be marketed rather like lottery tickets, employing lowincome people to sell tickets on a commission basis. Unlike standard lotteries, however, all ticket holders would win a prize in a disaster year, bot no prize would be given in non-disaster years. There is no need to restrict the insurance to farm families, and all types of rural households might find it attractive. This is because a decline in farm incomes in drought or flood years usually leads to a sharp contraction in the rural non-farm economy, and the incomes of many workers and businessmen decline in tandem. Drought insurance faces the same covariability problem as regional yield insurance, but because it is limited to a specific weather peril, it might be much easier to obtain international reinsurance. This prospect could be enhanced if the scheme were run by a commercial bank or insurer within the country.

Schemes of this kind have yet to be tried, and without pilot projects it will be difficult to assess their potential value. Their cost-effectiveness would also need to be compared with alternative means of assisting vulnerable households, such as relief employment schemes, and targeted food rations or income transfers.

# 2

# Sustainable Agriculture

## AGRICULTURAL DEVELOPMENT IN SUSTAINABILITY

Agricultural development has three distinct but related dimensions: the physical-technical, the economic-financial, and the institution-human. The physical-technical dimension addresses land utilization, agricultural technologies, research and extension, agricultural inputs, farm-to-market access, productivity and production maximization, and so on. Agricultural development from the economic-financial perspective is concerned with costs, factors of production, terms if trade, pricing policies, subsidies, incentives, credit, return on investment, market mechanisms, and the like.

The institutional-human dimension looks at knowledge and skills, organization and management, training, implementation capacity, social relations, politics, communication, motivation, participation, local government, public-private sector linkages, culture and values, historical experience, and so on. Agricultural development, as with much of the development field, has tended to be the province of the technical and economic specialists. Their language and concerns have dominated the analysis of rural and agricultural sector issues, the specification of agricultural development objectives, and the design of programmes and projects to achieve those objectives. The institutional-human dimension has frequently been either completely ignored or treated as a source of problems or constraints to achieving technical and economic targets, a residual category for anything not defined as technical or economic.

For example, a World Bank staff member notes that, "between 1976 and 1980 the Bank invested U.S. $920 million in agricultural projects without funds in any one of them being allocated to the staff training needed for the new activities on which those projects depended".

Recent work by Ruttan represents one effort to build theoretical integration of the three dimensions. On the applied front, the USAID-supported farming systems approach to agricultural development is one example of an explicit attempt to integrate the three dimensions operationally.

Recognition of the importance of institutional and human factors in socio-economic development in general has increased substantially over the past 15 years or so, and is now a high priority area of attention by donor agencies and developing country governments alike. Being a social technology, institutional development does suffer in the eyes of crop scientists, agricultural economists, and macroeconomic analysts from a lack of specificity, predictability, and hard- and fast rules.

Despite its "softness," institutional development is a frequently espoused need of developing country officials and an aim often expressed by donor agencies. U.S. foreign assistance policy, for example, aims officially "to build and maintain the social and economic institutions necessary to achieving self-sustaining growth".

This goal of institutional development is nowhere more important than in the rural and agricultural sectors, which continue to be the source of most employment in developing countries. Except for emergency relief, where the delivery of food for consumption is unavoidable, donor policies stress that assistance for the countryside should be an investment, that it generate a flow of benefits that endures after the external funds run out.

The cumulative lessons of experience show that to have lasting impact, donor efforts to improve rural areas in the developing countries, and agriculture in particular, need to work with durable indigenous institutions both at the central and local levels. It is not surprising that, in the 1990s, institutional sustainability is emerging as one of the major problem areas for sectoral management, as well as for public administration of core government functions in developing nations.

This chapter introduces the topic of institutional sustainability, focusing in particular on the agricultural and rural sectors, though its observations apply equally to other parts of the economy. To set the stage for the rest of the book, we start by briefly exploring the meanings of the key terms "sustainability" and "institution." Then we review the latest quantitative studies of sustainability, putting them in a comparative perspective.

The broad topic of development "sustainability" is obviously not a new issue. The folklore of foreign aid is replete with tales of equipment that never worked in the field and was left to rust, of well-intended recommendations that local people disregarded, of organizations that succumbed to apathy — if they ever functioned in the first place.

Critics of development assistance have always complained about this sort of waste. It has often been easier to be clear about the failed and was not sustained than to search out what worked, what lasted, and why. In the mid- to late 1980s, however, sustainability has surfaced as a distinct and pressing concern of policy-makers in donor agencies and in the developing world.

As the last decade of the twentieth century gets underway, the expansion of cropped area and the adoption of modern farming techniques are putting new, perhaps irreversible, stress on the natural environment in developing countries, while soft international markets for agricultural commodities are forcing reappraisal of the possibilities for the Third World to increase its farm exports. Conventional development strategies seem more and more to have run their course, yet fresh approaches have been difficult to find.

"Aid weariness" among the donor countries and high levels of indebtedness among the recipients have constrained the search for creative solutions. Finding ways to yes both external and indigenous development resources more efficiently to achieve broader and more lasting impact has become, accordingly, more difficult and more imperative.

Each of the academic disciplines cuts into sustainability problems from a different angle. Three points of view stand out, each reflective of the three dimensions of agricultural development previously noted. Agricultural and natural scientists tend to frame the problem of sustainability in terms of the long-term impact of current farming practices on the global resource base.

The chemicals and intensive cultivation associated with green revolution technology can degrade the environment and waste energy resources; scientific breeding of seed may reduce the native genetic diversity of crops, making food supplies more vulnerable to plant disease and pests. Because of such problems, agriculturalists and naturalistst are increasingly concerned with identifying and promoting methods of cultivation that can protect and enhance nature's assets, while continuing to produce enough food to satisfy burgeoning populations.

Thus the blue-ribbon Bruntland Commission has defined sustainable development as "development that meets the needs of future generations without compromising the ability of future generations to meet their own needs". Countries such as Nepal and Haiti are living (or dying) examples of unsustainable development, and represent cautionary stories for those who would ignore the ecological costs of detrimental policies and practices.

A downward spiral of events can result, with deforestation and erosion leading to food insecurity, extremes of poverty, class warfare, and ultimately to societal breakdown. Sub-Saharan African nations also appear to be in danger of proceeding down this road.

Economists, by contrast, are somewhat less preoccupied with the ecological dimensions of sustainability per se, since the mining of natural resources can be justified from an economic point of view. Their concern is not conservation by itself, but the appropriate tradeoffs between economic growth and environmental preservation.

Accordingly, economists tend to be troubled by the manmade market distortions that curtail the growth of production and employment in the Third World. The most widespread examples of these distortions are the ill-considered of politically convenient price regimens that penalize the rural sector in many developing countries. There are extended economic costs to such policies.

As the World Bank notes: "Discrimination against agriculture on a sustained basis not only reallocates resources within agriculture but also draws them out of it. As labor and capital move out and technical progress slow, the longterm losses can be large."

Management experts tend to make the development project or the organization their unit of analysis, and to worry about sustainability in terms of how to meet recurrent costs or permanently improve capacity for implementation. Development project management has emphasized achieving planned targets within a specified time frame and budget, thus responding more to donor agency needs for accountability than to recipient country capacity-building requirements.

Donor-supported projects, with their autonomous implementation units, frequently experience a collapsing "balloon effect" when external funds and advisers are withdrawn. Because they are often organizationally "quarantined" and protected, with special budget and staff, such projects may leave little lasting imprint on regular government agencies, and in many cases may actually weaken them by drawing off the best national talent.

The management perspective on sustainability, however, has evolved away from a narrow project focus, expanding to concentrate more on programmes integrated into developing country agencies and on the question of how to maintain an ongoing flow of goods and services to intended beneficiaries. From the administrative point of view, sustainability means building organizational capacity to perform over an extended time period (longer than the life of-project horizon) and assuring the continuation of useful and valued outcomes for client groups.

Specialists concerned with nonagricultural sectors cut into sustainability issues in their own way. Health experts, for example, are usually more concerned with the clientele than the implementing agency; they see the problem in terms of promoting affordable preventative health measures, a opposed to curative care alone, measures that enable poor people to enhance their physical well-being over the long run.

But the fundamental problems of sustainability in health and other fields are closely parallel to those facing farmers and rural administrators. The thread that connects the agro-environmental, economic, and management perspective on sustainability is effective social institutions, which much be in place to deal with all the current challenges to development.

Overcoming the environmental damage of modern farming, for instance, is likely to require (among other things) that scientific organizations attack novel research problems, that private enterprise produce greater quantities of more sophisticated farm inputs, that farm advisory services disperse new knowledge, that local authorities mobilize farmers to change collective and individual behaviour, and that farmers' organizations collect and market unfamiliar crops.

Even something as seemingly straightforward as price as price reform requires considerable institutional capacity to analyze and monitor the impact of new prices, let alone to manage any political backlash. The institutional dimension, for example, has emerged as critical to the success of the structural adjustment programmes and sector adjustment loans provided by the International Monetary Fund and the World Bank.

In virtually all problems of "development sustainability" the institutional dimension plays a key role. Unfortunately, many Third World countries lack the depth of institutions to carry through the complex and exacting duties likely to be needed to tackle the development tasks of the 1990s.

## Rule-oriented Institution

What are institutions? The concept is subtle and thus subject to confusion. According to Huntington definition, they are stable, valued, recurring patterns of behaviour. Institutions thus include rules or procedures that shaped how people act, and roles or organizations that have attained special status or legitimacy. Their importance is now being rediscovered by the social sciences, including the development subfield.

An example of a rule-oriented institution is a system of land tenure; whereas a role oriented institution could be the legal authority established to adjudicate disputes arising out of the land-tenure system. Both rules and roles can be institutionalized, the former as codes of law or custom, the latter as concrete organizations.

It is useful to keep in mind the distinction between these two types of institutions. Development assistance for agriculture and rural development is sometimes aimed at altering a village society's fundamental rules, for instance, by promoting tenure reform and land redistribution. When practitioners seek to build better institutions, however, the role-oriented, organizational definition is usually what they have in mind.

The USAID-sponsored institution-building literature of the 1960s, for instance, converged around organizations makes this matter of official policy, arguing that "because institutions become tangible only through the policies and actions of particular organizations, much of [our] institutional development effort will be focused on improving the policies and procedures of key organizations."

Most other donors and developing countries adhere to a similar definition, often narrowing it to mean formal collectives in the public sector, such as ministries, parastatals, agencies, or commissions. This book follows these conventions part way, by focusing on institutions in the organizational sense; however, our framework of analysis allows, indeed requires us, to simultaneously consider the effect on organizations of established codes of behaviour.

The development community's focus on role-oriented, organized activities unfortunately introduces another point of confusion, since "institutions" in this sense can be used loosely to refer to any formal or semi-formal collective entity. But not all organizations are institutions, any more than all institutions are organizations. As Perrow points out: "Some organizations are merely... rational tools in which there is little personal investment and which can be cast aside without regret.

Others become institutionalized. They take on a distinctive character; they become prized in and of themselves, not merely for the goods or services they grind out. People build their lives around them, identify with them, become dependent on them." Many rural organizations fail this test. When one refers to them as institutions it is often to speak hopefully about what they might become, and not realistically about what they, in fact, are.

A prime example of this failure is the experience of many developing countries with agricultural cooperatives, where the gap between what the coops were intended to be, and what they actually are, has been wide.

Our contributors concentrate on formally-constituted collectivities, but without prejudging whether they deserve the appellation "institution" in a strict sense. Further, the fact that the case studies tend to neglect informal modes of social organization, should not be interpreted as indicating that such institutions are less important in development. Nor should the focus of the case studies on discrete institutions be taken to mean that one may ignore the contextual factors within which organizations functions. To understand sustainability, it is important to keep this larger environment clearly in view.

## DIVERSE AND DIVERGENT CONCEPTS

Recent years have witnessed an immense outpouring of published literature under the general heading of 'Sustainable Agriculture'. In preparing a report on "Sustainable agriculture and food security" for the 1992 UN Conference on Environment and Development a scientific Commission of the International Council of Scientific Unions (ICSU) reviewed some 300 related publications [ICSU-CASAFA 1991]. The Commission reported how 'sustainable agriculture' is treated from a diversity of scientific, pseudo-scientific, biological, ecological, ideological and philosophical perspectives. The following are a sample of relevant significant publications referred to in the Commission's report: Agriculture Canada [1989], Altieri [1983], Boeringer [1980], Brown [1987], Brundtland [1987b], Bunting [1987], CGIAR/TAC [1989], Edwards et al [1990], FAO [1984], Harwood [1990], Jain [1983], NRC/NAS [1989], NABC [1989], Rodale [1983], Reganold and Papendick [1990], Swaminathan and Sinha [1986]. During the succeeding decade many more have been added.

As defined by the Brundtland panel, a sustainable agricultural system manages essential resources so as to satisfy the needs of all people presently dependent without compromising the needs of future generations. It is when authors try to prescribe agricultural production systems claimed to be of global ecological adaptability that confusions and contradictions arise. Geological and anthropological evidence indicate that, from the time of creation, the earth, its environment and inhabitant creatures have changed, sometimes gradually and evolutionary, sometimes in extreme and disruptive patterns. Many scientists contend that human activities are accelerating environmental and climatic change, some attributable to developments in agricultural technologies. The ICSU-CASAFA report illustrates how the relevant literature presents divergent opinions and recommendations based on experience with different locations, ecologies, technologies, environmental, climatic, social and economic circumstances. Sustainable systems are prescribed by commentators of variable international experience, some seemingly motivated more by ideological persuasions than from a logical, systematic scientific perspective.

Systems and technologies proposed for any farming community should be assessed in relevant context and require a comprehensive analytical understanding of the existing technologies, resources, constraints, prevailing ecological and climatic conditions; farmers' individual and collective capacities to tolerate the inevitable risk of adopting unfamiliar crop types or systems of cultivation. Farmers who produce surplus to their families' subsistence needs must have timely access to market information: the state of supply and demand, quantities and qualities demanded, current competitive prices, seasonal and unanticipated fluctuations in demand.

The ICSU-CASAFA report describes cultivable land as a fast-shrinking resource, and presses the need to raise food crop yields and farm productivity throughout Asia, Africa and Latin America. The report specifies the following urgencies:

a. to protect and enhance the natural resource base: farm land, surface and ground water, genetic diversity;
b. to reclaim and restore fertility in land degraded through misuse;
c. to promote systematic integration of production with post-production activities and to stimulate establishment and maintenance of rural agro-industries;
d. to devote more research to land capacity classification, soil and water management, to ecologically conservative land use.

The report illustrates the diversity, lack of consistency and many contradictions among definitions of, and prescriptions for, sustainable agriculture. It states: "It is unhelpful to propose any particular production system as sustainable in all ecologies, locations and societies. No specific system is suitable and sustainable for all circumstances ... Donor agencies should not seek to impose on poor farmers in developing nations systems designed to remedy wasteful practices in more affluent agricultural economies." The report contends that donors assign greater emphasis to agricultural production technologies than to post-production systems; to policy, social and economic considerations. The report urges far greater support for post-production systems; for technologies, logistics and economics of preservation, processing, transport, distribution and utilisation.

## Other Commentaries and Recommendations

A publication commissioned for the 1995 World Food Summit cites various proposed prescriptions and purposes for sustainable agriculture [Hulse 1995]:

a. to generate agricultural products adequate and acceptable in quantity, variety and quality;
b. to maintain environments favourable to humans and other organisms;

c. to prevent pollution of surface and ground water; to protect wild life and animal rights;

d. to prevent despoilment and degradation of fertile land by erosion, urban spread and activities unfavourable to agriculture;

e. to establish and maintain rural infrastructures essential to agricultural production and marketing;

f. to protect natural ecosystems and ensure long-term conservation over short-term exploitation;

g. to promote nutrient recycling and ensure a balance between immediate use and long-term stability.

Some observers, mainly from developing nations, insist that 'sustainable' is not synonymous with 'low-input' systems that reject all chemical inputs. Nutrients derived solely from animal manure and composted biological waste may be insufficient to maintain fertility in poor soils, to support increased crop production needed to feed expanding populations. Modern technologies must serve to complement not replace traditional practices [Jain 1983].

Based upon his extensive international experience, Harwood [1990] presents an impressive historical review of proposed sustainable agricultural systems. He suggests the word 'sustainable' implies a steady state to be realised only by defining long-term goals. He states there are almost as many opinions about sustainable production as there are authors, many opinions being based on too little hard data and little understanding of variability in resources. He suggests many recommendations are simple and devoted to single issues rather than total complex systems. He considers it futile to seek any widely applicable definition of sustainable agriculture and proposes the following as a workable description of 'sustainable': "Agriculture that evolves indefinitely toward greater human utility, more efficient resource use, an environmental balance favourable to both humans and other organisms." Harwood deplores the excessive use of resources and lack of concern for ecologies evident in intensive production systems practised by many American farmers.

## Conservation and Preservation

As a general guide to what is sustainable, the following are deserving of particular attention [Swaminathan 2001].

- *Soil health*: physical, chemical, microbiological properties and susceptibility to erosion.
- *Water quality*: water to be used for irrigation must be of low salt concentrations.

- *Plant health:* crops must be protected from insects, other pests, pathogens and competing weeds. Pests can be persistent year round in tropical climates.
- *Genetic homogeneity*: genetically homogeneous crop types cultivated over large areas are susceptible to pests and pathogens that eventually overcome a plant's natural resistances.
- *Abiotic stresses*: soil salinity, flooding, water logging, periodic drought, chemical contamination are serious impediments to crop cultivation.
- *Post-harvest management*: uniformity in maturity, appearance, post-harvest stability during storage and transportation are properties neglected by governments and many donor agencies despite their greater importance to urban consumers and processing industries.

All harvested crops, livestock and fisheries are in varying degrees perishable, critical conditions being active moisture content and ambient temperature. In perishable produce, deterioration starts soon after harvest and in tropical climates progresses rapidly if preservation processes are not applied. As urban communities expand, as spatial and temporal distances between rural producers and urban consumers increase, the probability of post-production degradation increases. Primary processing close to where perishable crops are cultivated serves to reduce avoidable waste and to provide off-farm employment for rural people.

Swaminathan refers to the potentially adverse consequences of climate change, the need for genetic selections that reduce dependence on chemical inputs, research to devise remediation methods to reduce soil and water pollution, for genetic modifications to transfer such useful traits as tolerance to salinity from mangrove plants to cultivated food crops. Tolerance to soil salinity would extend Asian land areas suitable for crop cultivation.

## Agricultural Resources

Resources essential to agriculture may be classified as

*Internal*: Resources that exist and are renewable within a farm's confines;

*External*: Resources purchased, contracted for, and imported from outside the farm.

Resources can be sub-classed as a. natural, b. synthetic and manufactured, c. socio-economic, d. human [ICSU-CASAFA 1991].

## Internal Resources

Land classification and soil characteristics are critical determinants of practicable farming systems, systems influenced by a. land type and profile, whether flat, sloping or undulating, naturally wet or dry, with or without drainage; b. soil properties: structure, depth, density, chemical and organic composition, acidity, water retention and permeability. Efficient land management restricts erosion and structural degradation; maintains fertility and inhibits loss of plant nutrients. Prudent water management restricts pollution by agricultural chemicals and organic effluents, conserves surface, fossil and ground water, prevents overdrafting of rechargeable aquifers.

*Internal flora* include a. natural vegetation, some useful as forage or fuel; b. cultivated crops; c. soil and air-borne microorganisms, some that symbiotically stimulate plant growth, others that are harmful pathogens.

*Internal fauna* comprise a. natural wild life: i. productive organisms: insect pollinators, predators and parasitoids that attack crop pests, earthworms that improve soil properties; and ii. destructive organisms: arthropods and rodent pests, some as vectors for pathogens; b. livestock husbanded for food, fibre and on-farm work.

## External Resources

Whenever the load on carrying capacity exceeds the internal resources, external supplementary resources become necessary. External resources may be purchased outright or obtained under contact. External resources can be employed to modify land profiles, construct water catchments, plant trees, and provide machines and equipment for irrigation and other functions. External power from electricity or fossil fuels may be needed to drive machines; agricultural chemicals derived from fossil fuels provide fertilisers, costs being influenced by the world price of oil. Use of agricultural chemicals has persistently increased; excessive use can be hazardous to human health, to other organisms and to environmental stability. Bio-fertilisers, bio-pesticides and biological control of pests reduce dependence on agricultural chemicals [ICSU CASAFA 1991].

## Agricultural Land and Soil

Soil is a living system: its structure, composition and biological diversity must be understood, protected and prudently managed if productive agriculture is to be sustainable [Brown and Wolf 1984]. The proportion of available arable land already cultivated varies from about 85 per cent across Asia to 22 per cent in sub-Saharan Africa.

However, in Africa, land classed as arable includes large areas of tropical forest, extensive humid ecologies heavily infested with tsetse fly, the vector for the pathogens that cause sleeping sickness in humans and anaemic debilitations in cattle. Tropical forest soils are fragile and rapidly eroded when deprived of tree cover. Effects of soil degradation include debasement of physical structure; erosion by wind and water; depletion of organic matter and natural nutrients; toxicity from salinity, alkalinity, acidity and chemical contaminants. Much soil degradation results from avoidable, human carelessness. [ICSU-CASAFA 1991].

Of the world's total ice-free land, an area of roughly 13.4 billion hectares, the potential arable land is barely 3.2 bn ha [24 per cent of the total] of which about 60 per cent is of low productivity, susceptible to many stresses that inflict degradation. It is estimated every year some 20 million ha become uneconomic for farming, and an additional 6m ha are degraded beyond practical reclamation [IBSRAM 1991].

An international assessment of soil degradation reports that more than 1.9 bn ha of world soils have suffered some form of degradation, ranging from light to extreme, most resulting from mismanagement, deforestation, over-grazing, over-exploitation and industrial chemical pollution. Of the Planet's degraded soils, 38 per cent are in Asia, 25 per cent in Africa, 15 per cent in Central and South America, 11 per cent in Europe (including eastern Europe) and 5 per cent in North America [ISRIC 1991].

More than 110m ha of Indian soils are significantly degraded: 81 per cent by wind and water erosion, 13 per cent from loss of fertility, 4 per cent from salinity, 3 per cent from water logging. The area degraded represents about 30 per cent of India's arable land [Swaminathan 1992]. Of Canada's territorial area of 10 million square km, less than one per cent is high quality arable land. It is this land, best for productive agriculture, located in Canada's most benign climatic zones, that is most favoured for urban and industrial development. The World Resources Institute estimates more than 50 per cent of the world's arable land has been severely degraded [WRI 2004].

Agricultural systems must be sufficiently sustainable to provide for the needs of inexorably expanding populations, with greatest growth in urban communities while rural producers proportionately decline. However defined, by purpose or practice, sustainable agriculture is wholly dependent on fertile arable land and safe water, both disastrously degraded, wasted and misused. Arable land and unpolluted water are fast becoming the main determinants of future sustainable agriculture.

## VARIATIONS ON THE SUSTAINABLE THEME

Much of the literature on sustainable agriculture originates in affluent industrialised nations and relates to intensive highly mechanised production systems that utilise substantial quantities of synthetic chemical inputs. Observers critical of subsidised intensive systems recognise in their nations less urgency to raise agricultural produce than to restrict disruption and degradation of natural ecosystems and resources. Though 'sustainable agriculture' is widely favoured, the following are other designations with reputedly similar objectives that have emerged in recent years.

### Alternative Agriculture

Literally interpreted, alternative agriculture describes any production system that differs from what is commonly practised. In some instances it is used to imply ecologically benign systems less dependent on agro-chemicals, other industrial inputs, more reliant on nutrient cycling, biological nitrogen fixation, propagation of natural enemies of crop pests and pathogens, conservation of soil and water quality, of biodiversity, protection of valuable symbiotic relations and biological interactions among crops, pests and pest predators.

### Biodynamic

These favour systems that use compost and humus to enhance soil structure and fertility.

### Ecological, Eco-biological, Socio-ecological

These are stated to be based on agricultural ecology harmonised with human ecology, to be protective of environments, to discourage use of chemical fertilisers, herbicides and pesticides.

### Low-input

Advocates propose least possible use of agro-chemicals, fossil fuel energy, mechanisation, commercial certified seed and pedigree livestock. They recommend on-farm breeding and seed multiplication, composting of farm waste and crop residues, animal and green manure. Interpreted literally, low-input systems would revert to primitive labour intensive closed farming systems.

### Low-input Sustainable [LISA]

Proponents assert that productivity is raised by more efficient use of certified seed, energy, mechanisation and other off-farm inputs; by soil and water conservation, conservative tillage, multiple cropping mixed with livestock, systematic recycling of manure and crop residues.

Others designated 'Regenerative' and 'Permaculture' appear similar to 'Biodynamic' and 'Organic Agriculture'.

## Organic Farming

Strictly speaking, all agricultural production is 'organic' in that every product of agriculture is composed of organic substances. Among the earliest proponents of organic farming were two Britons, Sir Albert Howard and Lady Eve Balfour. Before World War II Howard advocated adoption of production systems that used no synthetic fertilisers or pesticides. For more than 30 years, on her research farm, Lady Balfour conducted experiments in which conventional British farming systems were compared with 'organic' systems free from agro-chemicals. Rodale, the American high priest of organic farming, may have received his inspiration from the published work of Howard and Lady Balfour [Rodale 1983].

Food products from organic farming represent a relatively small but slowly growing proportion of foods purchased in Europe and North America. Across the EU, the agricultural area devoted to organic farming has increased from 0.5 per cent in 1993 to *ca* 3 per cent in the year 2000. The value of organic foods sold in the United States is estimated at $10 billion, across the EU at 8 bn Euro. The total value of retailed processed foods in the United States is over $500 billion, in the EU more than 350 billion Euro.

The *Codex Alimentarius* guidelines for organic farming specify the following objectives:

- to enhance biological diversity;
- to increase biological activity and maintain fertility in soils;
- to restore soil nutrients by recycling organic wastes;
- to rely on local renewable rather than imported manufactured resources;
- to promote healthy use of soil, water and air;
- to minimise pollution from agricultural practices.

Despite the interest of environmentally conscious consumers, rational development of organic farming has been hampered by confusion among concepts and definitions, comparable to diverse interpretations about 'sustainable agriculture'. A comprehensive review and state of progress among EU countries has been published [ECDGA 2000]. Another European publication estimates sales of organically produced foods among EU members at about 1 per cent of total food sales. It also comments on the higher prices for organic foods charged in supermarkets. In France price differentials between organic and non-organic produce range from 10 to 40 per cent for dairy products, close to 50 per cent for fresh fruits and vegetables [EEC 2002]. In an 89-page document the European Council presents regulations that specify permitted production, labelling and inspection practices for organic foods produced and sold in EU markets [EEC 1991].

The general principles of organic farming are unquestionably admirable: harmonisation of agriculture with nature, protection of ecologies and environments, conservation of soil fertility and water resources, recycling of wastes, humane husbandry of livestock. At present organic foods are purchased by a small proportion of dedicated consumers who can afford the higher prices. Prices may gradually fall if and when demand and production increase. Whether or not strict adherence to EU regulations, total prohibition of all synthetic inputs and transgenic crops can ensure sustainable food security for all the poorest nations and communities is open to question.

## Chemical Fertilisers

While those who claim to practise or encourage organic farming vary in specific recommended practices, most urge systems more environmentally benign, relatively free from synthetic pesticides, growth hormones and antibiotics. Whether chemical fertilisers should be permitted is not consistent among organic farming advocates. Fertilisers provide nitrogen, phosphorus, potassium and other elements to soils naturally deficient, or to replace soil nutrients depleted by growing crops. Chemical fertilisation began with importation of guano from sea birds in Peru and nitrates from deposits in Chile.

In addition to the main constituents – carbon, hydrogen and oxygen – 13 elements are known as essential plant nutrients largely taken up through root systems. Most essential nutrients had their origins in the earth's crust from which soils were formed by weathering when nutrients were released, dissolved in soil moisture, then absorbed by the root systems of wild and cultivated plants. The chemical composition of the earth's crust is highly diverse across the Planet; soils in different locations vary significantly in plant nutrient content (ICSU-CASAFA 1991).

Chemical fertilisers are compounded to provide nutrients in which soils are naturally deficient or depleted by cultivated crops. Nitrogen, essential for all plant growth, originates in the 75,000 tonnes of the gas which float above every hectare of the earth's surface. Elemental atmospheric nitrogen cannot be taken up by plants directly but, by complex biochemical reactions and various organisms, must be converted to ammonia or soluble nitrates. Microorganisms known as Rhizobia live in root nodules of leguminous plants where they convert atmospheric nitrogen into ammonia to fertilise the host plant. In industrial processes, elemental nitrogen is 'fixed' to produce chemical fertilisers in a form that release nitrogen from soil as and when needed by cultivated crops.

Substantial energy is absorbed in conversions of nitrogen into ammonia. In a microbial conversion, energy liberated from fourteen moles of adenine triphosphate [ATP] is required to produce one mole of ammonia. In a Haber-Bosch industrial reactor, pressures of *ca* 20 million Newtons/$m^2$ at 500°C energise the combination of nitrogen with hydrogen. Some nitrogen-fixing bacteria obtain their energy from organic materials in soils; in legume root nodules Rhizobia derive their energy from the host plant in return for which they supply the plant with accessible nitrogen. As Rhizobia fix nitrogen in legume root nodules, therefore, some organic farmers rotate or intercrop legumes with cereals that lack the physiological and biochemical mechanisms to fix nitrogen. There is hope that biochemical capability to fix atmospheric nitrogen may eventually be transgenically implanted into cereal crops.

## Other Sources of Nitrogen

Other nitrogen sources include animal manure; bio-fertilisers such as leaf mulch; compost from crop residues, in some instances assisted by earthworms as agents of conversion; green manure from leguminous plants and trees; such aquatic organisms as azola and blue-green algae. Manure, compost, mulch and other bio-fertilisers provide organic and fibrous materials beneficial to soil structure. Nutrient concentrations vary and rates of release from bio-fertilisers are not easily controlled, being influenced by soil temperature and moisture content. Release of nitrogen is generally slower from animal manure and composted residues high in lignin than from chemical fertilisers.

It is difficult to collect, store, distribute and utilise animal manure without significant nitrogen loss resulting from liquid run-off, leaching by rain, and volatilisation of ammonia. Manure stored in the open air in humid tropics with high rainfall can lose 70 per cent of nitrogen in 24 hours. Manure from cattle driven by pastoralists in semi-arid regions quickly dries and is blown away. Where wood is scarce and kerosene is expensive, animal manure may provide the cheapest source of household fuel. In contrast, chemical fertilisers are stable until used, designed and compounded to release nutrients as specifically needed in desirable composition and concentration.

Nevertheless, excessive application of chemical fertilisers may cause soil degradation, toxicity to earthworms and other useful soil organisms, and by leaching into ground and surface waters may cause insult to human health and harm to fish stocks.

In northern European countries, where high cattle stocking densities are common, serious nitrification of ground water results from manure accumulations that freeze during the winter and release high concentrations of water-soluble ammonia during the Spring thaw. Farmyard manure is a source of various coliform microorganisms some of which are pathogenic. An outbreak of listeriosis in Canada was traced to raw salad vegetables fertilised with farmyard manure.

## Contentious Issues

Closed farming systems with crops entirely sustained by recycled nutrients from within the farm perimeter are appealing and deserving of encouragement. FAO estimates that agricultural production must grow by at least 3 per cent annually over the foreseeable future if all the world's people are to be adequately fed. It is doubtful if closed farming systems applied worldwide can come close to realising FAO's objective.

A defect frequently encountered in the 'sustainable' literature is failure to define specifically what or who is to be sustained. Is 'sustainable agriculture' to be interpreted as maintaining a constant level of agricultural productivity, or does it entail production consistently sufficient to sustain an expanding, diversifying population? The 1990s across the world, and in several regions, population grew faster than cereal grain production [WRI 2004].

***1990 to 2000 per cent change***

| | Cereal production | Average yield | Population | Fertiliser use |
|---|---|---|---|---|
| World | +9 | +15 | +16 | [-3] |
| Asia | +19 | +13 | +19 | +35 |
| Africa | +18 | +6 | +31 | [-5] |
| China | +10 | +16 | +15 | +42 |
| India | +21 | +21 | +25 | +66 |

In at least ten African nations local cereal production decreased during the 1990s, a matter of serious concern since cereals provide the main source of food energy for the world's poorest people. The International Food Policy Research Institute predicts the demand for cereals across all developing nations will rise by about 50 per cent between 1997 and 2020 [IFPRI 2002].

The IFPRI publication forecasts that between 1997 and 2020 average cereal yields in developing regions will drop from 1.9 to 1.2 per cent/year; demand for meat will increase by more than 90 per cent. Nutritionally, cereals are utilised more efficiently when eaten directly by humans than when fed to animals to produce meat for humans.

Unless all livestock raised for meat are fed from natural pastures, food resources would be more conservatively utilised if all humans were vegetarians! IFPRI makes the gloomy prediction that by 2020, both China and India, the world's two largest nations that house almost one-third of all humankind, will become net importers of cereals [IFPRI 2002].

## SUSTAINABLE AGRICULTURE: STATIC OR DYNAMIC?

Agriculture cannot be considered sustainable if the quantity and quality of its products are inadequate to sustain the healthy survival of all who are dependent. Sustainable agriculture must be assessed by dynamic not static criteria. Simple maintenance of the *status quo ante* cannot ensure food security for expanding populations. From reliable assessments it appears that more than 5 million pre-school children throughout the world die every year from maladies caused or aggravated by malnutrition [IFPRI 2002]. Many causes of the melancholy state of, and forecasts for, future food and agriculture are offered:

a. a selfish indisposition of the affluent, both among and within nations, to be more conservative in their consumption habits, to share essential resources more equitably;
b. politicians who, to retain power and influence, invest more in armaments than to sustain more efficient systems of food and agricultural production and distribution;
c. sects more intent on inflicting their particular ideologies on the rest of mankind than in helping to alleviate poverty, social and economic inequities;
d. self-proclaimed authorities who advocate a specific set of production principles or a production system sustainable for the whole world.

### Sustainable Agriculture in the United Kingdom

A report submitted to the British Minister for Environment, Food and Rural Affairs by a policy commission on the future of farming and food in Britain begins with a 'Vision' of desirable objectives and conditions for food and agriculture [PCFFF 2002].

a. British farming and food sectors must be profitable, able to compete internationally, be good stewards of the environment, and provide a healthy, nutritious diet accessible to consumers in all income groups.

b. Production, processing and marketing must be holistically integrated, with more effective inter-communication among them.

c. The government is responsible for food safety, nutrition policy, animal welfare and environmental protection.

d. Public policy must recognise the interdependence of town and country.

The Report then compares its idealistic 'Vision' with the 'Present Situation' and states:

a. Farming and food industries [in Britain] are on a path that cannot be sustained over the long term.

b. Because of subsidies and price supports imposed by the EU Common Agricultural Policy, British and EU consumers pay more for food than the rest of the world, inflicting particular hardship on poor consumers.

c. The rural environment has been damaged by years of intensive agricultural production; while some farmers have adopted environmentally friendly practices, many cannot afford to do so.

d. Though the 'Vision' emphasises more holistic integration, significant disconnection exists among farm producers, food processors and retailers; unsatisfactory inter-sector relations and communications are detrimental to overall sustainable efficiency.

e. British consumers are uneasy about the wholesomeness and safety of the foods they purchase.

f. many health problems among Britons result from people eating too much of the wrong kinds of foods.

g. farming has declined in importance as an economic activity, farmers' share of retail prices having fallen. The total food and agribusiness industry contributes about 8 per cent, farming contributes less than 1 per cent to GDP.

h. farming is a highly competitive activity; farmers must develop more sensitive communications with consumers.

i. over 90 per cent of foods are purchased from supermarkets, allowing them undue influence over farmers and consumers.

j. the government must accept responsibility to support more economically efficient, sustainable farming and food systems; the government's role is to facilitate essential change, not to finance stagnation and protection of the status quo.

All of the issues raised in the Vision and Present Situation are discussed in considerable detail throughout the report.

## Sustainable Agriculture Initiative [SAI]

A few years ago the SAI was launched by the Unilever and Nestlé corporations, since when some 20 food processing and related agribusiness companies have become participants. Though there is sharing of general information, each participating company pursues its sustainable agriculture research and development independently with particular attention to raw materials essential to its brand products.

The quality of every processed food is critically dependent on the biochemical and biophysical properties of the raw materials from which it is processed. The properties of each raw material result from properties genetically inherited, degree of maturity at harvest, post-harvest conditions of handling, storage, transportation and the processing technologies applied in food factories.

The SAI programme seeks to establish and maintain close cooperation between farm producers and processors: to improve and, where necessary, modify farming practices to ensure delivery to processors of raw materials of defined acceptable properties. SAI encourages farming practices conservative of critical resources, protective of land, water, the environment, safety and wholesomeness of all raw materials. Listed among SAI objectives are:

a. to reduce risk of harmful residues such as pesticides, hormones and obnoxious contaminants;
b. to ensure safe on-farm working conditions;
c. to improve and expand farmers' knowledge and skills;
d. to preserve soil fertility, minimise erosion, encourage minimum tillage;
e. to promote farming systems that minimise waste and avoid pollution of surrounding environments;
f. to reduce use of chemical pesticides by adoption of biological and biopesticide control;
g. to use chemical fertilisers conservatively and effectively;
h. to conserve and protect scarce water resources;
i. to minimise energy consumed from non-renewable resources;
j. to explore practical economic alternatives to fossil fuels;
k. to protect biodiversity and wildlife habitats, the health and well-being of all farm livestock;
l. to employ environmentally benign production systems;
m. to raise farm productivity and profitability, to increase crop yields and reduce in-field and post-harvest loss;

n. to devise efficient farming systems based on systematic analysis and simulation modelling of all existing and potential biological, technological, social and economic parameters.

The primary objective of the SAI is to ensure long-term availability of agricultural raw materials essential to their industrial processes, materials that provide more than 2/3 of all raw materials used by these industries. SAI participants' dedication to the above objectives is highly commendable, motivated by the need to restrict extensive degradation of arable land; recognising the competition for land from urban and industrial spread, and the desirability of protecting biodiversity, nature reserves and wild life habitats.

SAI seeks to respond to the worldwide shortage of fresh water of which agriculture's consumption is close to 80 per cent of total world supply. SAI recognises growing competition from industries and urban communities, pollution of ground and surface waters by agricultural chemicals, municipal and industrial effluents, acid rain caused by industrial toxic emissions.

Guided by overall 'sustainable' objectives, SAI participants' R and D focuses primarily on agricultural crops and livestock of specific importance to their companies' processes.

Unilever's concentration is on tea, palm oil, various vegetables and fisheries; Nestlé gives precedence to milk, coffee, sugar and cocoa; Danone to cereals and fodder crops.

Environmental protection and resource conservation are also reflected in SAI participants' factory operations, where they pursue energy and water conservation, reduction and recycling of packaging and waste, elimination of refrigerants and other substances potentially harmful to atmospheric and environmental protection and conservation.

Though much maligned by 'anti-globalisation' activists, many transnational food and agribusiness corporations act more responsibly in conserving critical resources, protecting biodiversity, fragile ecologies and ecosystems, in improving agricultural practices and rural economies than can be said of some governments whose subsidies to farmers, damage to ecosystems and natural habitats in pursuit of oil and timber are destructive of ecologies and biodiversity.

Governments and development agencies could benefit by following the SAI pursuit of economic, holistic integration of all logistical activities and processes that constitute a benign integrated food system that begins with sustainable on-farm production and ends with delivery of safe, nutritious foods to consumers.

## INTERNATIONAL AGRICULTURAL RESEARCH: THE CGIAR

### Genesis and Progress

A history of the foundation and early years of the Consultative Group on International Agricultural Research (CGIAR) was published by the World Bank [Baum 1986]. Other publications provide details of the growth and activities of the CGIAR and the family of International Agricultural Research Centres it supports [ICSU-CASAFA 1991, Hulse 1995, CGIAR 2002]. Annual Reports of the CGIAR and individual IARCs are available from the CGIAR Secretariat at the World Bank in Washington.

All of the CGIAR research centres protest their dedicated commitment to sustainable agricultural systems and practices [CGIAR 1995; 2002]. But it is not always clear whether the primary purpose is for each centre to be permanently sustained or to ensure that national agricultural organisations, the developing country farmers and the populations they serve are to be sustained.

The CGIAR had its genesis in agricultural research sponsored by the Rockefeller and Ford Foundations. In 1943, Rockefeller began a bean and cereal improvement programme in Mexico, which eventually brought forth semi-dwarf high-yielding wheat phenotypes and, in 1963, metamorphosed into the International Wheat and Maize Improvement Centre (CIMMYT). In 1960 the two Foundations established the International Rice Research Institute (IRRI) in the Philippines, and later two centres to study tropical agriculture, one in Nigeria, and a second in Colombia. The CGIAR began in 1971 as an informal consortium of donor and development agencies with no legal constitution. From the start the World Bank has provided a supporting secretariat. Scientific assessments and recommendations come from the Technical Advisory Committee (TAC), a changing international group of experienced scientists with a secretariat housed at FAO in Rome.

In 1971 eleven members pledged $15 million to support the four existing IARCs. By 1976 membership had grown to 26 donors, who contributed $63 million to the four original and seven new IARCs. By 1992, CGIAR has expanded to a membership of 40 with a budget close to $300 million to maintain a family network of 16 IARCs and related programmes. The *CGIAR Report* for 2002 lists 62 donor members, including the three sponsors: World Bank, FAO and UNDP, and states there are now 16 IARCs that employ more than 8500 scientists and support staff. (A private communication indicates that in 2004 there were 15 centres and programmes, supported by over $400 million from 63 CGIAR members). The *2002 Report* refers to five priorities declared at the 2002

Johannesburg Conference: water, energy, health, agriculture, and biodiversity. Following a brief description of activities in each of the IARCs, the report describes Challenge Programmes in which several IARCS cooperate with emphasis on a. water and food, designed to increase productivity of water for food and livelihoods; b. bio-fortified crops for improved human nutrition: breeding crops with higher than normal contents of such micro-nutrients as vitamin A, iron and zinc. The report also describes a Global Conservation Trust, jointly sponsored by the CGIAR and FAO, to ensure diversity among the most important food crops through an endowment to provide funds sufficient to maintain crop genetic resources held in gene banks around the world. It mentions a report published by the International Peace Research Institute in Oslo that states that between 1989 and 1997 most of the major military and civil conflicts around the world occurred in regions heavily dependent on agriculture [CGIAR 2002].

A working group convened at FAO in 1989 by the CGIAR Technical Advisory Committee, with representatives from IARCs and several donor agencies, concluded that, "whereas agricultural production and farming systems have been well served by the CGIAR family, post-production systems, including breeding for functional properties essential to food processors, have been largely ignored."

## CGIAR: Original Purposes and Progress

At the outset CGIAR and TAC agreed that the IARCs' research would be devoted to improvement of crops which provide most of the food for developing nations: crops such as rice, wheat, maize, sorghum, several grain legumes, cassava, potatoes and other root crops. Later IARCs were created for research on livestock health and husbandry, fisheries, forestry, agro-forestry, bananas, and a centre for collection and classification of plant genetic resources and the conservation of biodiversity. Each IARC responsible for one or more food crops maintains an extensive germ-plasm bank of broad genetic diversity. National agricultural research and development programmes can select and receive genotypes and phenotypes of crop species suited to their needs and ecologies. Each centre provides training in the disciplines and activities related to its research programme. The CGIAR and its IARCs have assembled the world's largest, most diverse collection of plant genetic resources, amounting to some 750,000 classified accessions.

The early advances at IRRI and CIMMYT, much publicised as the 'Green Revolution', resulted from Mendelian plant breeding which shifted synthesised biomass from vegetative plant parts into the edible seeds. Plant breeders also developed and propagated rice types that matured more rapidly than farmers' traditional varieties. Rice farmers were thus able to plant and harvest two or three rice crops on the same land every year. During the 1960s and 1970s, rice production across Asia more than doubled; India increased its wheat harvest from 11 million tonnes in 1961 to 70mt in 2002 [IEDRC 2003].

Much early criticism complained that those who benefited most from higher yields were large-scale not small-scale producers with limited resources. The world's history of technological progress illustrates that innovative and novel technologies are first adapted and adopted by individuals and industries with access to resources sufficient to tolerate the inevitable risk of investing in unfamiliar practices. Eventually, the higher yielding crops and related production technologies were accepted by a majority of Asian rice farmers [Chandler 1982; Baum 1986; ICSU-CASAFA 1991; CGIAR 1995; Swaminathan 1999].

## Farming Systems Research

During the 1970s IRRI scientists laid the foundation for modern concepts of farming systems research, in a programme that began modestly with a few Filipino farmers but gradually spread to include many hundreds of rice farmers in 12 Asian nations. Western agricultural research traditionally began and continued on an experimental farm until promising results were eventually transmitted to farmers via extension services. In contrast, the IRRI farming systems research methods begin by assessing opportunities for, constraints and risks to increasing production by adaptation of new crop types and production technologies. The methodology starts with an analytical assessment of the biological, technological, physical, financial, economic and ecological resources existent among different farming communities; how and with what resources could more productive technologies be introduced with tolerable risk. The IRRI farming systems methodology has contributed significantly to increased crop production while conserving farmers' ecological and economic resources [IDRC 1981; Chandler 1982; Zandstra *et al* 1981].

## Estimated Benefits

Evenson and Gollin [2003] estimate results from what are variously designated 'high-yielding varieties' [HYV] and 'modern varieties' [MV] of food crops grown in developing countries. They compare changes in yield and total production during the 'Early Green Revolution (1961–80)' and the 'Late Green Revolution (1981–2000)' and the calculated influence of MVs mostly developed in CGIAR/IARCs. From 1961 to 1980 MVs accounted for 21 per cent of increased yield and 17 per cent of production increase; from 1981 to 2000, for 50 per cent of yield growth and 40 per cent of production increase. Most notable increases attributable to MVs were in Asia, the Middle East and Latin America. Little benefit from MVs was realised in sub-Saharan Africa where most increases resulted from expanding land under cultivation. Using a mathematical simulation model, the authors estimate that, without MVs in the period 1960 to 2000, food crop yields in developing countries would have fallen. The authors also claim that without MVs food prices would now be significantly higher throughout the world [Evenson & Gollin 2003].

### Subcontracting to National Programmes

An initiative that started in the International Centre for Research on Potatoes (CIP), and later pursued by IRRI, subcontracts programmes and projects to national agricultural research institutions in developing countries. The IARCs have trained thousands of scientists from poorer nations many of whom return to their native lands without funds and resources adequate to apply the knowledge and skills they have acquired. Subcontracting IARC programmes and projects to national research bodies serves a. to reduce operational costs, as developing country scientists cost less than their international counterparts; b. to increase resources and strengthen capabilities in national programmes. For agriculture to satisfy the needs of Africans, Asians and Latin Americans, resources and responsibilities should be systematically transferred from CGIAR/IARCs to national programmes and institutions.

## GENETIC MODIFICATIONS

### Genetic Modifications and Pest Control

All crops cultivated and wild are susceptible to attack by pests that infest, infect, predate or are parasitic. Most damaging are weeds and other competitive plants, arthropods (insects, mites, and spiders), pathogenic bacteria, fungi and viruses. Pests and predators cause insult and injury to vegetative and reproductive organs; pathogens inflict debilitating diseases; competitive and parasitic weeds deprive crops of essential nutrients; birds, rodents and insects wholly or partially destroy crops in the field and/or after harvest. Over countless millennia wild plants have evolved mechanisms and synthesised substances ('natural pesticides') by which to protect themselves from their enemies. Biophysical resistance mechanisms include tissues too tough or fibrous to permit penetration. Defensive biochemical substances are synthesised by growing plants: some toxic, some repellent to particular pests.

Cultivated plants and their pest enemies exist in dynamic relations. Pests mutate to overcome plants' natural resistance mechanisms and to tolerate chemical pesticides. To maintain resistance, plants gradually mutate to modify their effective defensive mechanisms. Being slower than insect and microbial mutations, it is among wild plants that stable evolved resistances are most evident. Cultivated plants can be protected by genetic modifications; by cultural practices such as intercropping where non-susceptible crops act as barriers to pest transmissions between same species; by biological control: propagation and release of organisms discovered to be enemies of particular pests; by chemical pesticides or bio-pesticides repellent or toxic to specific or a broad spectrum of pests. Protection by classical Mendelian breeding involves hybridisation of a susceptible genotype with a type resistant to particular pests. Modern transgenic techniques transfer and stabilise resistant traits from wild, distantly or unrelated organisms into cultivated species.

During the 1980s increases in food grain yields by classical Mendelian breeding were reaching practical limits; harvest yields were levelling off. In cooperation with the International Rice Research Institute, the Rockefeller Foundation developed and supported an international programme to devise techniques by which desirable genes could be transferred between distantly related or unrelated organisms. The Foundation acquired the active cooperation of an international group of experienced geneticists who worked to discover and devise innovative means of transferring genes between sexually incompatible species. The first objective was to enhance inheritable resistances in cultivated rice.

Of the 24 identified species of the *Oryza* genus, only two – *O.sativa* and *O.glabberima* – are cultivated as food crops. Over countless aeons the 22 wild *Oryza* species have evolved genetically inherited resistances to pests and pathogens to which cultivated species are susceptible. The objective of the RF-IRRI programme was to transfer genes that encoded for particular resistances from wild to cultivated *Oryza* species and thereby reduce rice farmers' dependence on chemical pesticides. Recognising that many persuasive activists suffer from extreme chemophobia, believing that any substance of biological origin is inherently less obnoxious and potentially harmful to human health than chemical synthetics, the RF/IRRI programme's objective of reducing reliance on chemical pesticides had logical reason.

Unfortunately, certain transnational agrochemical companies adapted the novel transgenic techniques to increase cultivated food crops' tolerance to chemical herbicides, thus permitting affluent farmers to apply higher than normal doses of herbicide to control competitive weeds. These commercial transgenic modifications were thus designed not to reduce but to increase demand for chemical pesticides.

Two international committees composed of experienced biological scientists pointed out that when a particular resistance is transgenically introduced into a cultivated plant, the plant's biochemistry is changed: it synthesises substances novel and foreign to its constitution, substances toxic or repellent to one or more pests. The target crop of the RF/IRRI programme was cultivated rice, the cereal that provides over 70 per cent of food energy for Asians. Much of the rice grown by small-scale Asian farmers is grown, harvested, distributed and eaten without scrutiny by a food regulatory agency.

Some years earlier, potato breeders in Britain and the United States had crossed standard cultivated varieties with Peruvian wild species to transfer resistance to certain pernicious pests. Fortunately, in both instances by happenstance, before the new hybrids were released to farmers and consumers, visiting scientists discovered that the resistances came from excessive and unsafe concentrations of a neurotoxic alkaloid. The new resistant types were immediately withdrawn and destroyed. With this experience in mind the two committees and a working group convened by the Canadian Food and Drug Directorate recommended to the rice breeders that, before release of any novel transgenic hybrids, biochemical analysts be engaged to determine the composition of the substances synthesised to provide pest resistance, to discover whether they were deposited in the edible seeds, or in vegetative plant parts. If in the edible seeds, toxicological and nutritional studies were essential to determine if concentrations present could result in insult or injury to the health of consumers. Had this advice not been totally ignored, subsequent opposition to GM crops might have been less clamorous, persuasive and restrictive of benefits that prudent genetic modifications could bring about [ICSU-CASAFA 1991, H & W Canada 1989, Hulse 1995].

## Adoption of GM Crops

A recent report by Clive James, a Canadian plant pathologist, states that in 2005, 54 million hectares (Mha) were planted to GM soybeans, 21.0 Mha to GM maize, 9.8 Mha to GM cotton and 4.6 Mha to GM rape-canola. GM crops are cultivated by roughly 8.5 million famers in 21 countries. The largest GM cultivators, in sequence of quantities produced, are the United States, Argentina, Brazil, Canada and China.

Disturbingly, James reports that 71 per cent of 90 Mha under GM crops are planted to herbicide-resistant genotypes of soybeans, maize, rape-canola and cotton. Though the original intention of the IRRI-Rockefeller GM programme was to reduce dependence on chemical pesticides, in fact the principal beneficiaries are herbicide manufacturers who also produce the herbicide-resistant seeds. The main active ingredient in the most extensively used herbicide is the isopropyl amine salt of glyphosphate. The potential toxicities of the active and other ingredients are still under study. In North America, the permitted residues of pesticides are scientifically prescribed and systematically monitored, but the same cannot be said of many poorer nations.

## Assessing the Safety of Genetically Modified Crops

Persuasive publicity opposed to all genetic modifications, particularly among Europeans, whose survival is not threatened by food shortages, could prove detrimental to future agricultural research and to realisation of food security among poorer nations.

While admitting that it is easier to discover acute or chronic toxicity than to demonstrate that a food or process can be assured as absolutely safe for all people under all circumstances, because of stringent food regulations, people resident among affluent nations have access to the safest food supply the world has ever known. But what may be demonstrated as 'safe' by reliable scientific analysis may not coincide with what is perceived to be safe by consumers influenced by strident propagandists.

The opportunities for crop improvement by molecular breeding and transgenic techniques are deserving of responsible scientific exploration. During the 1990s, it is estimated that more than 600 different chemical pesticides, valued at roughly $22 billion per year, were sold around the world [Dekker 1991]. Among nations of the EU, North America and Oceania, agricultural use of crop pesticides is rigorously controlled, while residues after harvest limited by law are systematically monitored. Among poorer nations, pesticide use is less stringently controlled and many instances of pesticide intoxications suffered by farmers and consumers have been reported. Broad-spectrum pesticides destroy not only targeted species but also many harmless and useful organisms. To induce inheritable resistances and thereby reduce use of dangerous pesticides is rational and sensible. Several scientifically reliable protocols to ensure safety of GM crops and derived foods have been published [Doyle and Persley 1996; IFBC 1990; IFST 1996; IFT 2000; IFT 2001; WHO 1991].

## Natural Pesticides

It has been estimated that some 5,000 'natural pesticides' – substances synthesised by plants to protect themselves – are present in foods sold in the United States. Though many have been identified, it is claimed that too little attention has been given to possible chronic toxicity, carcinogenicity, mutagenicity (disruption of DNA), and teratogenicity (induced congenital malformations) of many 'natural pesticides' [Ames 1983; Ames & Gold 1990]. An objective of many food plant breeding programmes was to eliminate or significantly reduce the concentrations of identified 'natural pesticides' known to be toxic, such as solanin the neurotoxin found in wild potato species. Cultivation of food crops deprived of their natural protection, require larger applications of chemical pesticides.

## GM Crops: Potential Benefits

A comprehensive review of opportunities for control of insect pests by molecular biology and transgenic modification concludes that these techniques will lead to improved pest management, facilitate precise studies of insect populations, genetic mapping, insect and crop resistance mechanisms. In addition to creating induced resistance in cultivated crops, genetic modifications of insects offer future means of biological pest control [Evans 1993].

Other potential benefits that genetic modifications of crops could provide include:

a. Greater tolerance to abiotic stresses – Tolerance to salinity can be transferred from mangrove plants into rice and other food crops [Swaminathan 2001]. Rice tolerant to soil salinity could substantially extend the land area across Asia for rice cultivation.
b. Enhanced nutritional quality – More desirable proportions of essential amino acids in cereal and legume proteins; higher contents of micro-nutrients deficient in many poor nations' diets [Swaminathan 2000, 2001].
c. Functional properties to expand suitability for food and other industrial processing – Scientists at the Plant Biotechnology Institute in Saskatoon, Canada, have modified the biochemical and biophysical properties of cereal starches to diversify their industrial applicability [Davis *et al* 2003; Demek *et al* 2000].

These are but a few of the possibilities for agronomic, nutritional and technological modification of food crops that could increase productivity and diversify utility of food crops for the benefit of poorer nations where, without significant improvement to protect and expand land suitable for cultivation, to raise harvest yields, and to reduce losses to pests and other degradations and debilitations, food security for many future millions in Asia, Africa and Latin America cannot be assured.

### Closing Comment

Practices and purposes deemed essential for 'sustainable agriculture' differ markedly among practitioners and commentators. As Harwood so cogently observed, there are as many prescriptions as there are authors. Commentators who live in affluent societies should be wary of prescribing remedies for poor folk who exist in fragile ecologies. The need to conserve biodiversity, arable land and water resources is undeniable. But no known crop or livestock production system is sustainable in all places and for all people. The IRRI farming systems methodology of acquiring a comprehensive understanding of what exists before offering remedies for improvement is a worthy example for all to follow.

Notwithstanding the many potential benefits to farmers and to food and health security offered by scrupulously designed and developed GM organisms, there is justifiable concern that the principal beneficiaries of agricultural GM are the producers of herbicides and herbicide-resistant food crops. Instead of reducing chemical pesticide use, herbicide applications are increasing.

In addition, it has been reported that some seed companies are experimenting with Genetic Use Restriction Technology (GURT), also known as 'Terminator' biotechnology. The plants are genetically modified to produce a sterile seed, so that farmers cannot save some of their harvested seed for later planting, but every year must buy new seed from the supplier. This practice is condemned under the Convention on Biodiversity formulated at the Rio conference, but there is evidence that some avaricious companies and the politicians they support and influence have hopes of ignoring the Convention.

## SUSTAINABLE FOOD SECURITY

### Diverse Definitions

As with 'sustainable development' there is no precise, universally accepted definition of 'food security'. In broad principle a state of food security exists where all individuals, families and communities enjoy consistent access to foods that in quantity, quality and biochemical composition provide hygienic, nutritional adequacy. The Brundtland Panel states: "Food security requires secure ownership of, or access to, food resources and income earning activities, including reserves and assets to offset risks, to ease shocks and meet all contingencies". This may be interpreted as all people having access to stocks and flows of food and cash sufficient to satisfy their basic nutritional needs. Everyone in the world need not be a subsistence farmer, but those who do not produce foods in quantities sufficient to satisfy their families' needs, must possess the means to acquire foods nutritionally adequate, readily accessible and affordable.

The report of the 1992 FAO/WHO international conference on nutrition states: "Food security is defined in its most basic form as access by all people at all times to the food needed for a healthy life.

Food security has three dimensions: assurance of a safe, nutritionally adequate food supply at national and household levels; a reasonable degree of stability in the supply of food throughout each year and from one year to the next; every household having physical, social and economic access to food sufficient to meet its needs."

While there is agreement in broad principle on requirements for food security, concepts diverge significantly in semasiological and biological specifics, they change over time, and differ among specialists and agencies as to what constitutes a nutritionally adequate diet. Non-scientific and quasi-scientific publications frequently fail to make clear distinction among 'hunger', 'malnutrition', 'famine' and 'starvation'; between 'chronic' and 'transitory hunger'. The noun 'security' derives from the Latin *'securus'* which means 'safe', 'free from worry and care'. People blessed with food security are safe from the risk of famine, starvation, chronic hunger and malnutrition. 'Famine' describes extreme food scarcity among large populations. 'Starvation' is derived from an Old English word meaning to die a slow lingering death from cold, disease, insufficient food and/or water. 'Malnutrition' defines a significant insufficiency or imbalance in essential nutrient intake [Latin *'nutrire'* = 'to nurse, suckle or feed'], 'sub-nutrition' sometimes describes people who have too little food. 'Hunger' describes distressing discomfort caused by want of food. 'Transitory hunger' is experienced from time to time by active persons who enjoy healthy appetites. 'Chronic hunger' describes a painful and debilitating condition caused by long deprivation of food sufficient to sustain healthy growth and activity [Hulse 1995]. The 1960s Freedom From Hunger Campaign, the 1974 World Congress' hope of eradicating hunger by 1984, and the FAO Millennium objective of reducing incidence of hunger by 50 per cent between 1990 and 2015, all relate to alleviation of chronic hunger and malnutrition.

Roughly 18 per cent of the world's population (800 million, including 200m children), 25 per cent of South Asians (300 million people) and 35 per cent of sub-Saharan Africans (190 million) are classified by UN agencies as undernourished: their caloric intake is lower than is required for healthy activity. In 1980 30 per cent of all people in LICs were considered undernourished. By the year 2000 this proportion had fallen to 18 per cent [FAO 2000].

## Food Security, Income and Poverty

In simplest terms, food security depends upon a dynamic balance among disposable income, demand, supply and distribution. Production of food crops, livestock and fisheries must progressively increase to satisfy expanding and diversifying needs and demands of growing populations. Efficient post-production systems, to ensure safe and effective protection, preservation, transformation and distribution must be economically integrated with production systems. During the early 1960s, affluent OECD nations were home to 34 per cent of world population and controlled close to 90 per cent of global GNP. By 1990, the affluent minority had fallen to 16 per cent of the total but controlled more than 82 per cent of global GNP. Between 1960 and 1990 a new group of middle-income newly industrialised countries emerged, most in East Asia and Latin America, a change with considerable implications for the Planet's economy and future patterns of food demand. In 1990 Asians represented 52 per cent, in 2002 some 59 per cent of world population. Among Asian nations, middle-income communities are growing most rapidly, their diets diversifying with dilating demand for livestock products.

Food insecurity, chronic hunger and under-nutrition are both dismal consequences and clear indicators of extreme poverty. People with money, power and control over negotiable assets rarely suffer from chronic hunger. Government politicians and military generals who suffer from chronic hunger are not a common spectacle around the world. Alleviation of food insecurity requires greater opportunities for paid employment and more equitable access to critical resources and assets among and within nations.

## Changing and Conflicting Concepts

A respected observer writes: "In the 1950s ... it was assumed that adequate agricultural production would assure access to adequate food in the market and in the household. In the 1970s it became clear that production and availability alone do not lead to food security, since those who lack purchasing power (from permanent employment income) lack access to a balanced diet. Now it is evident that where availability and access are satisfactory, necessary biological absorption of food requires clean drinking water, environmental hygiene, primary health care and education" [Swaminathan 2001].

Based on these considerations, the M. S. Swaminathan Research Foundation and the UN World Food Programme have published a *Food Insecurity Atlas for Rural India* [Vepa and Bhavani 2001]. The atlas illustrates marked variability among Indian States in food availability, absorption of nutrients, vulnerability to chronic hunger and in sustainable agricultural production. The atlas describes such non-food factors as income earning opportunities, health care facilities, education, sanitation and environmental hygiene.

Due to climatic variability, years with too little or too much rain, damage and destruction by pests, parasites and pathogens, crop and livestock production inevitably change from year to year. To compensate for years of poor harvests governments and/or regional organisations should maintain reserve stocks to carry over surpluses from regions and seasons of abundance to those of scarcity. The cost of holding reserve stocks should be calculated as an insurance investment not an unwarranted expense.

Food security need not be synonymous with food self-sufficiency, a nation's ability to produce all food required by its population. For diverse reasons: climate, topography, soil and climatic conditions, many nations cannot be self-sufficient and must depend on food aid or imports.

Substantial reductions in food spoilage after harvest could be realised by establishing primary food preservation and protection facilities in rural areas where crops are cultivated and animals husbanded. Food saved from spoilage increases total food available and enhances food availability and security.

## Public Policy and Food Security

Estimates of 'Global food security', based on the quotient of total possible food production divided by world population, may be a diverting statistical exercise, but is of little practical value. Equitable access to the planet's resources, though highly desirable, never has been nor is ever likely to be a realisable objective. To state that over the next half-century world food production must exceed all that was produced since settled farming began 10,000 years ago is interesting but offers little practical guidance for long-term planning by national governments. Food security can only be assessed and assured for discrete nations, communities and families within communities, a process that is the responsibility of national and local governments. International agencies can help weak and inexperienced governments by defining criteria to be considered and by providing long-term development assistance.

During the 1930s, the League of Nations Committee on Nutrition reported: "In the past, movement towards better nutrition has been largely the result of unconscious, instinctive groping for a better, more abundant life. Now needed is planning for better nutrition policies based on (a) consumption: bringing essential food within reach of all people; (b) adequate supply: production and equitable distribution of all essential food sources" [League of Nations 1936–1937]. The report urges (a) recognition of nutrition policy as of primary national importance; (b) better, widespread education on human nutrition; (c) more equitable distribution of income: "it is the poorest people who are most nutritionally deprived".

A committee of the US National Research Council appears to agree: "An important cause of malnutrition is absence of policies and programmes that foster the best use of available food supplies ... Governments continually make decisions that affect nutritional status with little knowledge of nutritional consequences. Alleviation of hunger and malnutrition depend upon increasing the right kinds of food, reducing poverty and improving the stability of food supplies" [NRC/NAS 1977].

## Governmental Responsibilities

Publications that discuss food security are many, vast and voluminous. They suggest primary causes, and propose various courses of remedial action. The League of Nations reports are among many that insist that food security be of highest priority in all governmental policies. Reutlinger [1987] states that food insecurity results from inappropriate macro-economic policies, from local economic and political structures that restrict the ability of households and individuals to have access to foods sufficient to satisfy their needs. Food insecurity may be transitory, lasting for relatively short periods, or chronic and of long duration, a consequence of poor planning to ensure resources adequate to produce or purchase sufficient food. Transitory hunger frequently results from unstable food production systems, sharp rises in food prices, or depression of disposable family incomes. If extreme and extended, transitory hunger can exacerbate into chronic hunger and eventually into famine. Famines have ravaged even when food stocks were plentiful: the 1943 famine in Bengal was precipitated by the price of a limited food supply rising to levels unaffordable by poor people.

Sen [1987] relates food security to entitlements and ownerships; families and individuals must have adequate access and purchasing power. Food security depends not on estimates of national aggregate demand but on fair exchange in international trade of goods and services. The poor must inevitably rely on food aid provided by social programmes. Chronic hunger and malnutrition are less influenced by a nation's total food supply than the degree of equity with which available resources are distributed and made accessible. Food and agricultural policies to ensure food security for all citizens should be a high priority for every government. Malnourished people, too weak to work effectively, eventually impose a financial burden on the nation's economy [Sen 1987].

## Strategic Food Stocks

Though some observers argue that food stocks of grain are of value only to alleviate temporary food insecurity, the prime purpose of strategic food stocks is to provide emergency reserves against regional, national, local and seasonal shortages, to stabilise food prices and counteract fluctuations in supply and availability. Reserve food stocks are customarily controlled by government agencies, authorised to maintain quantities sufficient for defined time periods. Reserve stocks must be replenished when they fall below a prescribed safe minimum. Reserve stocks, essential in regions susceptible to frequent drought, provide short-term relief but cannot be considered a panacea for chronic food insecurity.

Santorum and Gray [1993] list factors to be weighed in determining minimum reserve stocks. Stocks stored on farms are generally sufficient only to satisfy farmers' needs from one harvest to the next. Countries within a region that cannot individually afford to maintain national reserves can collectively contribute to and draw upon strategically dispersed food stocks. During the 1980s the Southern African Development Community (SADC) created and maintained a regional food security programme with grain stocks upon which member nations could draw in time of need.

## Food Aid

Where many are too poor to buy adequate food, where climatic change, civil strife and other adversities inflict serious disruptions on food production and distribution, food aid may be essential to survival and avoidance of famine. The UN World Food Programme [WFP] is the most efficient distributor of food aid, designed to provide relief or to support development. Food aid for relief alleviates acute stress among people whose food supplies are disrupted by war, civil unrest, enforced migration, severe climatic catastrophes, and/or massive crop failure. Food aid for development provides food as wages for economically productive work and to support basic social services. Food aid for development may support labour-intensive agricultural production, construction of roads, rural infrastructures, amenities and services essential to developing economies. Food aid helps to stabilise food prices, to provide for pregnant women, nursing mothers, school feeding and pre-school infants.

WFP administers food aid most efficiently. Misdirected, ill-conceived bilateral food aid can do more harm than good. One could cite a weary litany of bilateral food aid where indiscriminate dumping of donors' surplus crops and livestock products have depressed prices and discouraged agricultural production in recipient countries. Food aid administered by many bilateral agencies is motivated more by the donor's political and commercial interests than relieving acute or chronic distress among malnourished people.

WFP resources have been seriously overstretched, because of exceptional demands resulting from military, civil and inter-ethnic strife and conflict, mass migrations such as in Darfur, cataclysmic disruptions as produced by the Tsunami. As extreme demands for food aid will inevitably persist, more generous and sustained support for the World Food Programme should be a high priority for all donor agencies.

## Nutritional Adequacy

What constitutes nutritional adequacy has attracted scientific and unscientific interest for many centuries. Alchemists searched for but never discovered the *Elixir vitae*. In Britain during the mid-nineteenth century desirable dietary standards were described as: "the least costly foods to prevent starvation and disease during massive unemployment". The League of Nations standards were intended "to marry health and agriculture" [League of Nations 1936–1937]. Our most primitive ancestors, by trial and error, identified certain foods as pleasant and digestible, others as obnoxious, insulting or injurious to health, foods acceptable and foods unacceptable. Our early antecedents chose their foods from plants and animals indigenous to their ecologies and climates resulting in widely different food preferences among nations and communities.

## Dietary Standards and Recommendations

Estimates of human nutrient requirements were first codified during the latter half of the nineteenth century. From the time of their establishment, FAO and WHO have sponsored many meetings, committees and working groups to study and recommend desirable nutrient intakes, each meeting re-estimating recommended levels based on new knowledge. At least 50 national, regional and international agencies recommend what they consider desirable intakes of essential nutrients. Recommendations are customarily cited as specific amounts of each essential nutrient to be ingested daily. Some authorities name them Recommended Daily Allowances [RDAs], others Recommended Daily or Dietary Intakes [RDIs], others Recommended Nutrient Intakes [RNIs]. The evolution of recommended dietary standards was reviewed by several authors [Beaton 1991; Hulse 1995; IUNS/CABI 1983]. The purpose of RDAs is to encourage healthy eating habits and to prevent diseases attributable to nutrient deficiencies. Earliest motivations for government dietary standards arose in times of severe stress and distress; to feed nations at war or when suffering economic depressions, widespread poverty and malnutrition.

Recommended Daily Allowances [RDAs] the preferred American designation, are intended to serve several purposes: (a) to assess adequacy of food intake among selected population groups; (b) for national food and agricultural policy formulation; (c) to balance food nutritional quality with human need; (d) to provide adequate food during emergencies, particularly for the poor and disadvantaged; (e) to prescribe and enact standards for nutrient labelling of packaged foods. RDAs were first designed to meet the needs of the majority, often embodying margins of safety to allow for variable needs among people of different ages, activities and conditions of health. It is now customary to prescribe specific RDAs for citizens of different sexes, ages and levels of activity.

RDAs were first calculated by estimating the average physiological requirements for each absorbed nutrient, a value then modified to compensate for incomplete absorption and utilisation, variability among persons, bioavailability among different food sources. RDAs are based on foods consumed in normal diets, not on pure substances or concentrates, and take account of nutrient variability within and among particular foods. At high levels of intake some nutrients, such as vitamin A, may be toxic so that upper limits of safe intake are prescribed. It is difficult to prescribe precise RDAs for all conditions of men, women and children and their variable activities. Experiments on human beings are expensive, time consuming and only small groups of volunteer subjects can be studied in confined controlled conditions. Using humans as experimental animals also raises ethical objections. RDAs are intended for people as they go about their normal daily activities but food intakes and nutrient balance are difficult to determine where subjects move freely and are not confined in a metabolic research facility. Data based on *ex post facto* recall by subjects going about their normal lives are at best approximate.

Despite a vast accumulation of nutritional data, with more sensitive analytical tools available, nutrition remains a relatively imprecise and speculative science. Concepts change as more evidence comes to light. Thirty years ago it was believed military men working in Arctic climates needed more food than those working in temperate zones so Canadian troops in the Arctic north were given packaged rations that provided 5,000 kg Cals/day.

It was later realised that with insulated clothing and the need to move slowly to avoid heavy perspiration that could freeze, caloric requirements were no higher in cold than in temperate climates. Since caloric intake must balance energy expenditure, people who perform heavy work or strenuous exercise need higher caloric intakes than sedentary persons.

People who change from energy-intensive activities to a sedentary life-style tend to eat more than they need. Overeating with resultant obesity is disturbingly evident among the richer nations and affluent persons in less developed economies. It is reported that between 25 and 30 per cent of Europeans and North Americans are excessively obese, at risk of cardiovascular diseases, diabetes and other related disabilities. The obese among the affluent reflect Nerissa's comment in Shakespeare's *Merchant of Venice*: "They are as sick who surfeit with too much as those who starve with nothing."

## Nutritional Fashions

Responsible nutritional scientists seek to encourage rational eating habits based on the best available knowledge. The alchemist's pursuit of the *Elixir vitae* is far from moribund among publicists who proclaim all manner of diets guaranteed to ensure good health and longevity [Hulse 1995]. The penchant for nutritional panaceas has been evident for the past 50 years.

During the 1950s vitamins were in favour, in the 1960s proteins and amino acids were in fashion, from the 1970s cholesterol and cardiovascular dysfunctions obsessed public attention, next came the health benefits promised by dietary fibre. The present passion is for nutriceuticals, proclaimed in principle some 4,000 years ago by the Chinese, and for 'functional foods' (which beg the question: what are non-functional foods?), neither of which are specifically defined.

## Variability in Composition

The nutrient content of every food crop varies among genotypes and conditions of cultivation. Among the world's wheat collection on a dry weight basis, protein contents range from 6 to 23 per cent, lysine (as per cent of protein) from 2 per cent to 4.3 per cent [Hulse and Laing 1974]. A review of over 1,300 sorghum analyses reported protein contents (N × 5.7 DWB) to range from 7.1 to 14.2 per cent, lysine (mg/g nitrogen) from 71 to 212, iron (mg/100g) from 0.9 to 20.0, niacin [mg/100g] from 2.9 to 6.4. Significant variability was reported among samples of pearl millet (*Pennisetum typhoides*), chickpea (*Cicer arietinum*), faba bean (*Vicia faba*), and lentils (*Lens esculenta*) [Hulse 1991a, 1994, 1995; Hulse and Laing 1974; Hulse *et al* 1980].

Accurate chemical and biochemical analyses are difficult and demand professional competence and experience. Inexperienced technicians are unaware of many sources of error among standard analytical methods. Some 16 sources of error were identified in the classical Kjeldahl method of determining protein nitrogen [Williams 1974].

## Changing Recommendations

Since the nineteenth century, recommended daily intakes of all known nutrients have changed among and within responsible agencies. In 1985 a WHO committee recommended a daily protein intake of 0.6g per kg of body weight; a US NRC committee advocated 0.75 g per kg of body weight. How US RDAs have changed since first promulgated in the early 1940s.

***US NRC Recommended Daily Allowances***

| | 1945 | 1989 |
|---|---|---|
| Protein (g/day) | 100 | 59 |
| Ascorbic acid (mg/day) | 100 | 60 |
| Iron (mg/day) | 15 | 12 |

***UK Recommended Daily Nutrient Intake***

| | 1950 | 1979 | 1993 |
|---|---|---|---|
| Vitamin A (RE* mg) | 1500 | 750 | 800 |
| Ascorbic acid (mg) | 20 | 30 | 60 |
| Niacin (mg | 12 | 15 | 18 |
| Iron (mg) | 12 | 10 | 14 |

* RE = retinol equivalent

A comparison of daily intakes recommended by 39 governments and three international agencies, updated in 1995, illustrates many inconsistencies. Intakes prescribed for young active males (M), for teen-age women (F). Disparities of similar magnitude were evident for all other ages and levels of activity.

***Recommended nutrient intakes (RNIs) from 41 countries and agencies***

| | | Highest RNIs | Lowest RNIs |
|---|---|---|---|
| Energy (kcal) | M | 3200 | 2500 |
| | F | 3000 | 2030 |
| Protein (g) | M | 105 | 37 |
| | F | 96 | 31 |
| Vitamin A (mg RE) | M | 1500 | 600 |
| | F | 1500 | 450 |
| Vitamin D (mg) | M | 10 | 2.5 |
| | F | 10 | 2.5 |
| Ascorbic acid (mg) | M | 95 | 30 |
| | F | 85 | 25 |
| Iron (mg) | M | 24 | 5 |
| | F | 35 | 12 |

Many more people would be classed as malnourished if the highest rather than the lowest or some level in between were the international standard.

For international comparative purposes it is sensible to use the intakes prescribed by WHO and FAO. In light of these many differences and uncertainties, all RDAs, RDIs and RNIs should be regarded as indicative but not scientifically absolute. Disparities among nations may reflect different life styles and activities or various differences of opinion. A Canadian publication states:

"The primary use of nutrient recommendations is to form a basis from which to advise the public on how to choose a healthy diet. It would be a mistake to conclude that those who fail to adhere to each recommendation are doomed to suffer associated chronic deficiency disease, or that by strict adherence consumers will avoid deficiency diseases" [H and W Canada 1990].

## Recommended Daily Allowances (RDAs)

RDAs, RDIs and RNIs can be useful as general guides for cognisant officials responsible for formulating national health, food and agricultural policies. For most consumers, RDAs have little practical meaning. Consumers are better advised by illustrated guides to healthy eating based upon mixed diets of foods that are familiar, acceptable and seasonably available [H and W Canada 1992].

Human nutritional requirements, criteria used to arrive at recommended levels of intake, consequences of serious deficiencies, food intolerances and allergic responses are discussed in greater detail in other publications [Beaton 1991, IUNS/CABI 1983, Hulse 1995].

## Urban Growth and Food Security

Predictions of the size to which the world population will grow, and for how many persons the planet can provide sustainable food security 50 or 100 years from now, though intellectually diverting, are of little practical utility. One can estimate with reasonable certainty what could be produced with present resources. What may be feasible 50 or 100 years from now is indeterminable.

Of more pressing urgency are the consequences of rapidly changing demographics and rising disposable incomes among urban communities. More than half the people of Asia and Latin America live in large towns and cities, expanding in size and populations all over the world.

Though some food can be raised in city gardens and horticultural allotments, urban communities rely on food safely transported from rural farms, inland and maritime fisheries.

As people migrate from rural to urban communities and disposable incomes rise, eating habits change. Growing demand for processed convenience foods is illustrated by the replacement of cereals, pulses and cassava, hand ground and cooked into some form of porridge, with processed foods such as bread, biscuits, cakes and pasta. Of critical consequence for natural resource conservation is the growing demand for livestock products, which across SE Asia is rising by over 20 per cent a year.

Few of the world's cities were planned and laid out in expectation of the volume and density of vehicular traffic destined to grow progressively worse in all cities on all continents. To alleviate traffic congestion, the City of London Authority has imposed heavy tolls on private vehicles that enter the city on working days.

Too little investment is devoted to improving and maintaining rural infrastructures, to rural-to-urban and intra-urban transportation systems. The longer time taken to transport perishable food materials from rural producers to urban processors and markets, the greater is the probability of biological and economic loss.

Urban food security demands consistent access to a safe adequate supply of acceptable, affordable foods dependent on a. technologically efficient systems of preservation and processing, primary preservation being applied soon after harvest; b. logistically efficient and economic means of storage and transportation; c. reliable markets which retail foods that are safe, of acceptable quality, in unit quantities and prices that all segments of urban society can afford; d. national food and agricultural policies that provide and maintain the infrastructures, services, transportation facilities, social and economic conditions essential to efficient holistically integrated food systems.

## Final Observation

The Brundtland principle of providing adequately for present need without detriment to the needs of future generations has particular relevance for food security.

The future need, globally and among specific communities, and future agricultural capacity cannot be predicted, it is therefore, incumbent on all peoples, their governments, industries and institutions, to ensure that resources critical and essential to food production, preservation and distribution be protected and used conservatively. Degradation of arable land suitable for crop production must be arrested, where possible degraded land must be reclaimed. Water, the most critical and scarce of all essential resources, must be protected from pollution and misuse, whenever feasible purified and recycled.

In expectation of unrelenting inter-ethnic conflict, massive migrations, increasing severity of damage by hurricanes and cyclones, together with severe droughts, all consequent upon climate change, the need and demand for food aid is more likely to grow than contract. More generous and sustained support for the World Food Programme must be accepted as an obligation by all affluent nations and donor agencies.

Assurance of sustainable food security for expanding congested urban populations demands efficient integration of all activities and procedures between rural producers, food processors, urban markets and consumers. Urban food security in all nations calls for systematic planning by professionally qualified men and women of imaginative vision. For, as the *Book of Proverbs* so wisely observes: "Without vision the people will perish".

# 3

# Globalization on Indian Agriculture

Agriculture, since its invention and inception, has been the prime and foremost activity of every culture and civilization throughout the history of mankind. Agricultural practices secured the food supply. It also resulted in establishment of farming settlements that subsequently grew in villages, towns, cities, states, countries and empires.

The history of agriculture is broadly divided in four phases, differing widely in date according to the region.

The four phases are:

1. *Prehistoric Agriculture*- Believed to have started near 10,000 BC or earlier, it is considered as first prototype of agricultural modules. The main agricultural sites were in southeast and southwest Asia, in what are now India, Iran, Iraq, Israel and the adjoining areas. Nile River in Egypt, Danube in Europe and Indus-Ganga in India were the main center of the agricultural farmlands.

   This was the age of agricultural innovations as many new tools and practices were invented due to compulsion of planting, sustaining and harvesting the crops. Use of wooden and earthen tools as well as some domesticated animals are major highlights of this period.

2. *Historical Agriculture*: This age roughly covers the period from 2500 BC to 500 BC. It marks the introduction of metals in agricultural tool making. Agricultural sites meanwhile grew proficient and covered much of southeast Asia, middle-east, Europe, North America and eastern Asia.

   Mechanicals improvements in agricultural implements and development of storage and irrigation method marks the highlight of this age. For the first time, trade in agriculture also started, becoming the precursor to modern agribusiness industry.

3. *Feudal Agriculture*: The feudal age of agriculture ranges from 500 BC to 1700 AD, reaching its height near 1000 AD and then getting static thereafter.

   This period is chiefly characterized by expansion of agriculture and irrigation and consolidation of agricultural methods. Use of fertilizers, in form of natural manures to boost production was a major breakthrough of this period.

   The usual practice of this age was of intensive agricultural throughout Europe, middle east and southeastern Asia.

4. *Scientific Agriculture*: By 17th century AD world was seeing a rapid rise in its population. This was also the age of imperial expansions in Asia, Africa, Americas and oceanic countries.

   These led to exploration of new methods and practices in agriculture. It also resulted in increased attention on commercial form of agriculture that was based solely on profit than subsistence purposes. Agriculture trade assumed global proportions and many new economies thrived on specialized production of one or other agricultural products.

Industrial revolution also led to many new inventions for agricultural sector, particularly in irrigation, sowing, and harvesting techniques. But it was the modern inventions in biotechnology and genetics that have completely changed the form of agriculture production. New variants of seeds and plants are manufactured in laboratories and genetically modified to suit particular consumption and commercial needs. This has enabled the world to meet the food demands of for a population that is crossing 6.5 billion mark.

The liberalisation of India's economy was adopted by India in 1991. Facing a severe economic crisis, India approached the IMF for a loan, and the IMF granted what is called a 'structural adjustment' loan, which is a loan with certain conditions attached which relate to a structural change in the economy. The government ushered in a new era of economic reforms based on these conditions. These reforms (broadly called Liberalisation by the Indian media) can be broadly classified into three areas: Liberalisation, privatization and globalization. Essentially, the reforms sought to gradually phase out government control of the market (liberalisation), privatize public sector organizations (privatization), and reduce export subsidies and import barriers to enable free trade (globalization).

There was a considerable amount of debate in India at the time of the introduction of the reforms, it being a dramatic departure from the protectionist, socialist nature of the Indian economy up until then. However, reforms in the agricultural sector in particular came under severe criticism in the late 1990s, when 221 farmers in the south Indian state of Andhra Pradesh committed suicide. (The damage done, 2005) The trend was noticed in several other states, and the figure today, according to a leading journalist and activist, P. Sainath1, stands at 100,000 across the country. (Sainath, 2006) Coupled with this was a sharp drop in agricultural growth from 4.69% in 1991 to 2.06% in 1997.

This paper seeks to look into these and other similar negative trends in Indian agriculture today, and in analyzing the causes, will look at the extent to which liberalisation reforms have contributed to its current condition. It will look at supporting data from three Indian states which have been badly affected by the crisis: Andhra Pradesh, Maharashtra and Kerala. Andhra Pradesh's (AP's) experience is particularly critical in this debate because it was headed by Chief Minister Chandrababu Naidu, who pursued liberalization with enthusiasm. Hence liberalization in AP has been faster than other states, and the extent of its impact has been wider and deeper. (Sainath, 2005)

## Indian Agriculture Today: A Snapshot

Agriculture employs 60% of the Indian population today, yet it contributes only 20.6% to the GDP. (Isaac, 2005) Agricultural production fell by 12.6% in 2003, one of the sharpest drops in independent India's history. Agricultural growth slowed from 4.69% in 1991 to 2.6% in 1997-1998 and to 1.1% in 2002-2003. (Agricultural Statistics at a Glance, 2006) This slowdown in agriculture is in contrast to the 6% growth rate of the Indian economy for almost the whole of the past decade. Farmer suicides were 12% of the total suicides in the country in 2000, the highest ever in independent India's history. (Unofficial estimates put them as high as 100,000 across the country, while government estimates are much lower at 25,000.

This is largely because only those who hold the title of land in their names are considered farmers, and this ignores women farmers who rarely hold land titles, and other family members who run the farms.) (Sainath, P) Agricultural wages even today are \$1.5 - \$2.0 a day, some of the lowest in the world. (Issac, 2005) Institutional credit (or regulated credit) accounts for only 20% of credit taken among small and marginal farmers in rural areas, with the remaining being provided by private moneylenders who charge interest rates as high as 24% a month. (Sainath, 2005) An NSSO$_2$ survey in 2005 found that 66% of all farm households own less than one hectare of land. It also found that 48.6% of all farmer households are in debt. The same year, a report by the Commission of Farmer's welfare in Andhra Pradesh concluded that agriculture in the state was in 'an advanced stage of crisis', the most extreme manifestation of which was the rise in suicides among farmers. Given the performance of agriculture and figures of farmer suicides across the country, this can be said to apply to Indian agriculture as a whole.

## AGRICULTURAL DEVELOPMENT STRATEGIES

The lack of development in African countries has caused their economies to fall further and further behind those of the leading industrial nations. Many different development strategies have been tried. Some African countries have successfully encouraged investment in mining, tourism and industry. In agriculture, producers have been encouraged to move away from subsistence farming towards a more commercial approach as governments realised that income generated from the sale of surplus production could be used to improve productivity.

Agricultural development in ECA has faced an uphill struggle for the last twenty years. In an effort to stimulate development many countries borrowed heavily from bodies such as the IMF and from the commercial banking sector. These loans were not granted without strings attached, however. Most African countries were obliged to liberalise their economies by adopting significant policy changes often applied in packages known as Structural Adjustment Programmes (SAPs). These programmes included a number of elements but generally included requirement to:

- devalue the currency (to discourage imports and make exports more competitive),
- to make the currency freely convertible with other currencies,
- to cut public expenditure (in order to lower taxes),
- to dismantle state controlled marketing boards,
- to privatise state-owned industries (to raise capital and stimulate competition),
- to cut import restrictions (to encourage local industries to become more efficient),
- to allow foreign companies to freely repatriate profits (to encourage inward investment),
- and to boost exports.

The economists who designed SAPs were convinced that the only way African countries could transform their economies was to encourage inward investment and earn foreign exchange to invest in infrastructure and lay the foundations for industrialisation. These measures assumed that any country could compete in the world market if production and investment was concentrated in areas where they were deemed to have a competitive advantage.

The only activity in which ECA nations could be said to have a competitive advantage in the world market was in the production of agricultural products and the exploitation of natural resources such as forestry, fishing and mining. The major flaw in this strategy was that similar advice was given to almost all tropical countries at the same time. Coffee-producing countries were encouraged to boost coffee production; sugar producers should produce more sugar, and so on. This resulted in over-production of these commodities which caused prices to plunge in the international markets. On average, current prices of tropical products (taking dollar inflation into account) are only about one seventh of those prevailing in 1980 (UN General Assembly). Economists call this phenomenon the fallacy of composition - less income is earned as more commodities are produced.

Another component of SAPs which many observers believe to have been counter-productive was the requirement to cut public expenditure. All too often this meant a cut in health programmes, education and agricultural extension. These measures have tended to reduce, rather than enhance the flexibility of the workforce and to curtail agricultural development. Overall, the record of inward investment has been poor and the ending of currency controls has increased opportunities for transfer pricing abuse (where companies over-price imports and under-price exports to reduce tax liability).

The most important SAP reform affecting the distribution of agricultural products has been the dismantling of state-controlled marketing boards and the practice of setting fixed purchasing and sales prices for commodities. It was assumed that government control of markets had obscured the forces of competition in supply and demand in the economy. A free market system would unleash these forces and increase productivity. It would force producers to meet the demands of consumers both in price and quality. Farmers would be able to buy inputs cheaper from competing suppliers, and the country, as a whole, would become more competitive in world markets.

Unfortunately, competitive and transparent markets did not emerge spontaneously (Shepherd). Most African farmers have too little land to produce truck-loads of goods and they are widely dispersed over the countryside. There is not enough business to encourage more than one trader to operate in many areas. Farmers have no means of communicating with the outside world or even the nearest town and they are often unwilling to risk the investment of bringing their goods to market resulting in considerable waste. Laws may have been passed which ban collusion among traders to pay low prices to farmers and charge high prices to consumers, but there are often insufficient resources to enforce such laws. Most traders have no experience of free market conditions and are reluctant to put their fellow traders out of business with serious competition.

Advocates of SAPs point to examples of countries that have improved their economies after adopting SAPs (World Bank) but there are few in Africa. Most ECA countries were not able to implement SAPs until relatively recently but rates of poverty have increased in many of these countries. Intense conflict, both within and between countries of the region, drought, desertification and, now HIV/AIDS have further weakened economic development in ECA. Most critics of the reform process, however, acknowledge that markets in African countries must be made more competitive and SAPs are designed to do that but this process may take a considerable time.

### *Trade Agreements*

Economic links between ECA countries and their former colonial rulers have been maintained since independence. The economies of these countries have been moulded to meet the needs of their European counterpart for a hundred years or more and it would have been difficult for them to make the necessary changes in production patterns to trade successfully with other countries. The Europeans too needed to maintain supplies of raw materials and export markets in Africa and to protect the business of their trading companies.

In 1975 all ten countries covered by this study became party to the Lomé Convention. The Convention established trade, aid and cultural relationships between 15 European countries and 71 so called ACP (African, Caribbean and Pacific) countries which had either been colonies of, or had had strong historical links with, Europe. This agreement did not rule out bilateral or multilateral agreements with other countries but did give ACP countries preferential access to European markets. ECA countries have also decided to try to stimulate regional trade by bringing their economies closer together in regional economic agreements such as COMESA and SADC.

East Africans have exchanged goods and ideas with many other peoples of the world for millennia. In these exchanges of goods, cultural links have been established which have influenced ECA life at all levels – in religion, the arts, public sector structures, the economy and agriculture. In the last decade or two, however, this process has accelerated tremendously. There is no agreed definition of globalisation. It is simply a term which has been used recently to describe the impact of innovations in communication and transport systems on trade and the growing interdependence of nations due to economic sophistication and burgeoning output. In addition, high levels of protection between trading blocks of countries are breaking down as barriers to trade are reduced. These changes have made it possible to increase the volume of trade between countries in agricultural products.

It became clear that overall levels of trade could be increased if trade barriers were reduced, where there was agreement to do so, and that international trade should be governed by mutually agreed rules. The most active trading nations have been keen to find new markets for their goods and to reduce the barriers to free trade. These countries, however, have been reluctant to expose their own markets to foreign competition, especially unfair competition from subsidised or sub-standard goods.

At the international level, global liberalisation was stimulated by the General Agreement on Tariffs and Trade (GATT) which was first implemented in 1948 as a mechanism to promote free and fair trade among member countries. Several rounds of negotiations of trade rules have occurred throughout the history of GATT. The Uruguay Round, which began in 1986, was the eighth of the GATT rounds. In April 1994, officials from more than 100 countries gathered in Marrakech, Morocco to sign the Uruguay Agreement and to confer the role of further trade reforms on the newly established World Trade Organisation. The reform process is by no means complete. Almost all countries have now committed themselves to the objectives associated with their membership of the WTO. (6 out of the 10 countries covered in this study are WTO members). In order to meet these objectives, countries are obliged to further reform their existing internal economic and external trade policies. The future of trade and agriculture in ECA is inextricably linked to the rate and direction of these reforms.

## THE CRISIS FACING INDIAN AGRICULTURE

The biggest problem Indian agriculture faces today and the number one cause of farmer suicides is debt. Forcing farmers into a debt trap are soaring input costs, the plummeting price of produce and a lack of proper credit facilities, which makes farmers turn to private moneylenders who charge exorbitant rates of interest. In order to repay these debts, farmers borrow again and get caught in a debt trap. I will examine each one these causes which led to the current crisis in Andhra Pradesh, Kerala and Maharashtra, and analyse the role that liberalisation policies have played.

As was mentioned earlier, AP's experience is particularly relevant in this analysis because of its leadership. Let me explain in detail. Chandrababu Naidu, Chief Minister of Andhra Pradesh from 1995-2004, was an IT savvy neo-liberal, and believed that the way to lead Andhra Pradesh into the future was through technology and an IT revolution. His zeal led to the first ever state level (as opposed to national level) agreement with the World Bank, which entailed a loan of USD 830 million (AUD 1 billion) in exchange to a series of reforms in AP's industry and government. Naidu envisaged corporate style agriculture in AP, and implemented World Bank liberalisation policies with great enthusiasm and gusto. He drew severe criticism from opponents, saying he was using AP as a laboratory for extreme neo-liberal experiments. Hence, AP's experience with liberalization is critical.

## The Debt Trap and the Role of Liberalisation

### *The Debt Trap: High Input Costs*

**Seeds:** The biggest input for farmers is seeds. Before liberalisation, farmers across the country had access to seeds from state government institutions. For example, AP's APSSDC3 produced its own seeds, was responsible for their quality and price, and had a statutory duty to ensure seeds were supplied to all regions in the state, no matter how remote. The seed market was well regulated, and this ensured quality in privately sold seeds too. (The damage done, 2005) With liberalization, India's seed market was opened up to global agribusinesses like Monsanto, Cargill and Syn Genta. Also, following the deregulation guidelines of the IMF, 14 of the 24 units of the APSSDC's seed processing units were closed down in 2003, with similar closures in other states.

This hit farmers doubly hard: in an unregulated market, seed prices shot up, and fake seeds made an appearance in a big way. Seed cost per acre in 1991 was Rs. 70 (AUD 2) but in 2005, after the dismantling of APSSDC and other similar organizations, the price jumped to Rs. 1000 (AUD 28), a hike of 1428%, with the cost of genetically modified pest resistant seeds like Monsanto's BT Cotton costing Rs. 3200 or more per acre, (AUD 91) a hike of 3555%. (Sainath, 2005) BT Cotton is cotton seed that is genetically modified to resist pests, the success of which is disputed: farmers in Andhra Pradesh and Maharashtra now claim that yields are far lower than promised by Monsanto, and there are fears that pests are developing resistance to the seeds. (Menon & Jayaraman, 2002) Expecting high yields, farmers invest heavily in such seeds. Also BT Cotton and other new seeds guarantee a much lower germination rate of 65% as opposed to a 90% rate of state certified seeds.

Hence 35% of the farmer's investment in seeds is a waste. (Sainath, 2004) Output is not commensurate with the heavy investment in the seeds, and farmers are pushed into debt. The abundant availability of spurious seeds is another problem which leads to crop failures. Either tempted by their lower price, or unable to discern the difference, farmers invest heavily in these seeds, and again, low output pushes them into debt. Earlier, farmers could save a part of the harvest and use the seeds for the next cultivation, but some genetically modified seeds, known as Terminator, prevent harvested seeds from germinating, hence forcing the farmers to invest in them every season.

**Fertilizer and Pesticide:** One measure of the liberalisation policy which had an immediate adverse effect on farmers was the devaluation of the Indian Rupee in 1991 by 25% (an explicit condition of the IMF loan). Indian crops became very cheap and attractive in the global market, and led to an export drive. Farmers were encouraged to shift from growing a mixture of traditional crops to export oriented 'cash crops' like chilli, cotton and tobacco. (The damage done, 2005) These need far more inputs of pesticide, fertilizer and water than traditional crops. Liberalisation policies reduced pesticide subsidy (another explicit condition of the IMF agreement) by two thirds by 2000. Farmers in Maharashtra who spent Rs. 90 an acre (AUD 2.5) now spend between Rs. 1000 and 3000 (AUD 28.5 – 85) representing a hike of 1000% to 3333%.

Fertilizer prices have increased 300% (Sainath, 2005) Electricity tariffs have also been increased: in Andhra Pradesh tariff was increased 5 times between 1998 and 2003. (Seeds of ruin, 2005) Pre-liberalisation, subsidised electricity was a success, allowing farmers to keep costs of production low. These costs increased dramatically when farmers turned to cultivation of cash crops, needing more water, hence more water pumps and higher consumption of electricity. Andhra Pradesh being traditionally drought prone worsened the situation.

This caused huge, unsustainable losses for the Andhra Pradesh State Electricity Board, which increased tariffs. (This was initiated by Chandrababu Naidu in partnership with Britain's DFID4 and the World Bank.) Also, the fact that only 39% of India's cultivable land is irrigated makes cultivation of cash crops largely unviable, but export oriented liberalisation policies and seed companies looking for profits continue to push farmers in that direction. (Isaac, 2005)

## The Debt Trap: Low Price of Output

With a view to open India's markets, the liberalization reforms also withdrew tariffs and duties on imports, which protect and encourage domestic industry. By 2001, India completely removed restrictions on imports of almost 1,500 items including food. (The damage done, 2005) As a result, cheap imports flooded the market, pushing prices of crops like cotton and pepper down. Import tariffs on cotton now stand between 0 – 10%, encouraging imports into the country.

This excess supply of cotton in the market led cotton prices to crash more than 60% since 1995. As a result, most of the farmer suicides in Maharashtra were concentrated in the cotton belt till 2003 (after which paddy farmers followed the suicide trend). (Hardikar, 2006). Similarly, Kerala, which is world renowned for pepper, has suffered as a result of 0% duty on imports of pepper from SAARC5 countries. Pepper, which sold at Rs. 27,000 a quintal (AUD 771) in 1998, crashed to Rs. 5000 (AUD 142) in 2004, a decline of 81%. As a result, Indian exports of pepper fell 31% in 2003 from the previous year. (Sainath, 2005) Combined with this, drought and crop failure has hit the pepper farmers of Kerala hard, and have forced them into a debt trap. Close to 50% of suicides among Kerala's farmers have been in pepper producing districts. (Mohankumar & Sharma, 2006)

## The Debt Trap: Lack of Credit Facilities and Dependence on Private Money Lenders

In 1969, major Indian banks were nationalized, and priority was given to agrarian credit which was hitherto severely neglected. However, with liberalisation, efficiency being of utmost importance, such lending was deemed as being low-profit and inefficient, and credit extended to farmers was reduced dramatically, falling to 10.3% in 2001 against a recommended target of 18%. (Seeds of ruin, 2005) A lack of rural infrastructure deters private banks from setting up rural branches, with the responsibility falling on the government, which has reduced rural spending as a result of its liberalisation policies.

Rural development expenditure, which averaged 14.5% of GDP during 1985 – 1990 was reduced to 8% by 1998, and further to 6% since then.

This at a time when agriculture was going through a crisis proved disastrous for farmers, who turned to private money lenders who charge exorbitant rates of interest, sometimes up to 24% a month. (Seeds of Suicide, 2005) With input costs and output prices being what they are, coupled with crop failures and drought, they are pushed into debt which is impossible to repay. 12 out of India's 28 states have 50% and higher indebtedness among farm households.

Andhra Pradesh has the highest percentage of indebted farm households — 82%. 64.4% of Kerala's farm households and 54.8% of Maharashtra's farm households are indebted (NSSO, 2003) Indebtedness has been identified as the single major cause of suicides in both Andhra Pradesh, Kerala and Maharashtra. (Analysis of Farmer Suicides in Kerala, 2006, Who's suicide is it anyway, 2005, Saving small farmers, 2005)

## Liberalisation and How it Failed

Branco Milanovic, a World Bank economist describes how he believes liberalisation helps developing countries achieve growth: 'when a country lowers trade barriers, reduces government intervention in the market in order to allow market forces to operate freely, increases competition and attracts foreign investment, it will increase productivity and reduce inefficiency, which will lead to economic growth, and in a few generations, if not less, the poor will become rich, illiteracy will disappear, and poor countries will catch up with the rich.' (The damage done, 2005) This argument is an economic rationalist one, which views government intervention with profound suspicion, and has equally profound faith in unfettered market forces. (Whitwell quoting Robert Manne, 1998) What Mr. Milanovic neglects to mention, though, is that rich countries which now preach liberalisation protected their 'infant industries' at the time they began to industrialize, till they were strong enough to compete globally.

The US government, for example, had a protectionist trade policy in the late nineteenth century to help US companies become competitive in the world.

Besides, apart from wool, the US, Germany, Britain and France were all almost self-sufficient in the raw materials that they needed for industrialization, and took off from that platform, a luxury that India and other developing countries do not have. (Issac, 2005) As German economist Friedrich List says, the adoption of these values (of liberalisation) assumes that all countries are at the same starting place, which (as we have seen above) is clearly not the case. (Issac, 2005) In fact, it is this very reason that has brought about the crisis that Indian agriculture is facing today.

Most farmers in India were already in a position of minimum security, with no education system, credit facilities, access to alternative employment, or efficient technology. Their only support was government subsidy and regulation. Liberalisation policies came in and dismantled their only support structure. It halted the sharp reduction in rural poverty from 55% in the 1970s to 34% in the 1980s. Not only has the incidence of poverty in rural areas not gone lower than 34% in the 1990s, it has gone to higher levels of 42% in individual years. (Ghosh, 2000)

The second most popular argument of the economic rationalists in favour of liberalisation is that competition will weed out the inefficient, and in the growth that ensues, employment will be provided in other areas of the economy, thus lifting the poor out of poverty. This argument however assumes that the poor will be able to take advantage of the opportunities presented to them. As

Robert Issac says in 'The Globalization Gap', "Globalization encourages the well positioned to use tools of economics and politics to exploit market opportunities, boost technical productivity, and maximize short-term material interests." This is compounded in India, where the gap between one who is 'well positioned' and one who is not can be extreme. With a lack of investment, chances of generation of rural employment are slim.

Unemployment and underemployment are chronic problems in India, with the rate of unemployment being close to 10% in 2004. (Sainath, 2005) Primary education in rural areas is mismanaged and bad quality, and there is no system which helps agricultural workers find alternate employment, or develop alternate skills. (Chossudovskly, 1997) In the face of such obstacles, it is nearly impossible to expect agricultural workers to shift to alternate fields. Coming back to AP, the IT Revolution spearheaded by Chandrababu Naidu attracted companies like Google, Amazon, Microsoft and Dell and created thousands of jobs. However, given the skills and education of most farmers, it is obvious that none of this translated into job opportunities for them.

The final argument that supporters of globalization have is the much touted 10% reduction in poverty (60 million decline in poor) in India in the year 2000. However, this figure was challenged by experts. Poverty is defined according to how many people consume less than the nutritional minimum prescribed. (2400 calories for rural areas, and 2100 for urban areas) Major changes in survey design in 1999-2000 not only made the resultant estimates incomparable to previous years' estimates, but an over-estimation of consumption (meaning people were getting enough food, hence were not considered poor) meant a sharp reduction in poverty figures. After experts challenged it, the Planning Commission of India accepted that the figure was inaccurate, and could not be compared to previous years' estimates, hence the 10% drop in poverty is incorrect. With adjusted figures, experts have determined that the decrease in poverty was a mere 2.3%, and that the number of poor increased by nine million in 2002 as compared to 1999.

## Liberalization and 'Growth'

Many economists now concede that the relationship between liberalisation and growth are 'uncertain at best'. According to the Center for Economic and Policy research, which studied impact of liberalisation reforms on the developing world, key economic and social indicators such as increases in life expectancy, infant and child mortality, education and literacy levels slowed down in the 20 years between 1980 and 2000 when liberalisation policies were implemented, compared to the 20 years leading to 1980. (The damage done,

2005) This defeats the economic rationalist argument of free trade eliminating poverty, since the 20 years leading up to 1980 witnessed high protectionist policies and trade barriers. Following the suicides in 2000, the World Bank and Britain's DFID abandoned power reforms in Andhra Pradesh four years before schedule. It admitted that it had 'substantially underestimated' the 'complexity of the process' and that there must be 'increased consultation with the farmers to get their acceptance' of any further reform. (The damage done, 2005).

The Andhra Pradesh government sponsored report by the Commission of Farmer's Welfare squarely laid the blame for its agrarian crisis on the state and central government's policies: "While the causes of this crisis are complex and manifold, they are they are dominantly related to public policy. The economic strategy of the past decade at both central government and state government levels has systematically reduced the protection afforded to farmers and exposed them to market volatility and private profiteering without adequate regulation; has reduced critical forms of public expenditure; has destroyed important public institutions, and has not adequately generated other non-agricultural economic activities." A report on suicides in Kerala similarly held the liberalization policies of the government responsible. (Mohankumar & Sharma, 2006)

It is clear that the liberalisation policies adopted by the government of India played a dominant role in the agrarian crisis that is now being played out. However, this is not to say that privatisation, liberalisation and globalization are per say bad, or inherently inimical to an economy. It is the 'one size fits all' brand of liberalisation adopted by the IMF and the World Bank which forces countries to privatize, liberalise and globalize without exception which has failed. Without taking into account the state of an economy, and in this case, the state and nature of the agricultural sector in India, the IMF and the World Bank, with the cooperation of the Indian government, embarked on mismatched reforms, which have caused misery and despair among millions of Indian farmers, driving large numbers of them to suicide. It is also essential to break the link between aid and liberalisation, which caused India in the first place to accept the conditions of the IMF.

Remember that India was on the brink of a financial crisis in 1991 when it applied for the IMF loan and accepted its conditions—perhaps the course of economic reform in India would have taken a very different course if there was no urgent need to borrow from the IMF. The start to this process may have already occurred: recognizing the failure of its liberalisation policies, (and perhaps also the failure of DFID with AP's power reforms) the Blair government of Britain announced in 2004 that it will no longer make liberalisation and privatization conditions of aid. In another blow to the neo-liberal lobby, Chandrababu Naidu suffered the worst ever defeat in the 2004 state elections in his party's history, with rural AP clearly rejecting his brand of World Bank sponsored liberalisation. The battle, however, has not yet been won. It is essential for the rest of the G8 to follow Britain's example in order to influence World Bank and IMF policy towards India to ensure blind liberalization is not pursued, and so that countries like India can adopt tailor-made reforms to suit their economy.

## CIVIL SOCIETY & FARMERS' REPRESENTATION ON AGRICULTURE

"Organic agriculture is defined as a holistic food production management system, which promises and enhances agro-ecosystem health, including biodiversity, biological cycles and soil biological activity. It emphasizes the use of management practices in preference to the use of off-farm inputs, taking into account that regional conditions require locally adapted systems. This is accomplished by using, where possible, agronomic, biological and mechanical methods, as opposed to using synthetic materials to fulfill any specific function within the system." — FAO/WHO, Codex Alimentarius Commission

Agriculture in India is in crisis. The failure of the 'Green Revolution Technology' (GRT) is clearly evident in stagnant agricultural production, a mounting spiral of ecological problems, relentlessly rising input costs, and increasing farmer indebtedness and suicides. From the small, peasant farmer to the FAO, there is global consensus that the GRT path is unsustainable. It impoverishes both the farmer and the natural resource base of agriculture, and provides toxic food and water to consumers. The increasing billions of dollars spent every year to subsidize 3% of the population engaged in farming in the USA, is stark proof of its economic bankruptcy. Large corporates who provide agro inputs, are the real beneficiaries of chemical based agriculture.

India's 'National Commission on Farmers' recently reported: "40% of Indian farmers would like to leave farming if it is possible to do so". This summarizes the enormity of the present agricultural crisis and the challenge facing the nation: how to safeguard agricultural incomes or provide alternative livelihood support to a quarter billion people who are potential future economic and ecological refugees – uprooted by mounting farm production costs and a rapidly degrading natural resource base.

While some corporate interests are now lobbying for Genetically Modified (GM) species as the solution, a growing body of scientific testimony and evidence – from both India and round the world – warns against the many serious and irreversible dangers these pose. Such hazards include the uncontrolled and unwanted spread of certain genes and genetic traits; emergence of resistant and more virulent secondary pests; potential health hazards; and other unpredictable problems. Evidence of the economic counter-productivity of GM in the long run, is already emerging. In China, for instance, which has been using Bt cotton for over 7 years, there has been a severe rise of secondary pests when the bollworm is controlled, resulting in the same levels of pesticide spraying as before the use of Bt seeds. In India, there is an alarming record of an increase in crop failure, farmer suicides and deaths of grazing cattle after the use of Bt cotton in rainfed areas of Andhra Pradesh and Vidharbha, Maharashtra. The high cost, high risk and ecologically damaging GM technology is not the solution for our agrarian crisis, and will further damage Indian agriculture.

Both GRT and GM based agriculture require chemical intensive and fossil fuel intensive inputs, which directly contribute to greenhouse gases and climate change: an estimated 25% of the world's carbon dioxide emissions, 60% of methane gas emissions and 80% of nitrous oxide emissions. In fact nitrous oxide, produced by the use of nitrogenous fertilizers, is 200 times more potent than carbon dioxide as a greenhouse gas. Chemical based agriculture estimatedly uses ten times more energy as input than the calories it produces as food. Shifting to bio-diverse, organic systems can contribute significantly to mitigating global warming and climate change. National vulnerability to the rising costs of fossil fuels, inescapable with chemical farming, is also much reduced.

Small and marginal farmers, whose land holdings are below 2 ha, constitute almost 80% of all Indian farmers, and more than 90% of them are dependant on rain for their crops. With GRT, their costs of cultivation and risks of crop failure are so high that often the farmers cannot recover even the money spent. Between 1990-91 and 1995-96, chemical fertilizer costs increased by 113%, and pesticides by 90%, whereas the wholesale price of wheat went up only 58%. Minimum Support Prices for all crops, except sugar, were 38% to 50% lower than the actual cost of production. Per capita food grain production has fallen to levels lower than the 1939-44 famine. Despite the thousands of crores spent on fertilizer and other subsidies, farmers are increasingly in debt and despair. Apart from the necessary immediate reliefs in crisis areas, through write off of loans/interest, and appropriate Minimum Support Prices for agricultural produce, there is urgent need for a planned and vigorous promotion of low-cost, low-risk, high nutrition, holistic and sustainable farming systems to reinvigorate Indian agriculture, and to stem the rising tide of farmers' indebtedness, distress and suicides.

It is the experience of increasingly large numbers of farmers that holistic farming systems, based on scientifically proven techniques, are a very successful remedy to the economic and ecological crises engendered by GRT Holistic farming systems rely on available natural resources and rebuild the ecological capital on which all agriculture is dependent.

They also greatly reduce or totally eliminate the dependence of farmers on the purchase of expensive external inputs such as seeds, fertilizers, pesticides and herbicides. Particularly for rain fed farming and horticulture, which cover at least 60 \% of our cultivated land, holistic farming should be adopted on a large scale without fear of loss of production.

Depending on the extent of prior damage by chemical-intensive monocultures, there is some drop in productivity in the transition phase if adequate biological inputs are unavailable.

But within 2 to 3 years, the system is equally, or more, productive than other systems – and on an improving and sustainable growth path. This has been scientifically validated internationally by FAO reports, and by ICRISAT Patencheru (O P Rupela & others). Better nutrition (balanced and free of toxins), and distributive justice (helping the poorest) are additional important benefits of a holistic approach.

There is clear scientific empirical evidence that a holistic, organic approach to farming nurtures numerous effective microbes occurring in nature that make all the essential nutrients available to crops, and improve vital physical qualities of the soil, namely its porosity, aeration, moisture absorption, drainage, and resistance to erosion. Crop yields thus increase in a sustainable manner, input costs fall and farm incomes rise.

Mixed cropping – central to holistic farming – greatly increases the total yield of biomass, which in turn, absorbs more carbon dioxide and thus ameliorates global warming. It also helps maintain soil fertility through on-farm recycling, heightens security against climatic vagaries and crop failures, and improves the nutritional balance of local diets.

This will also help in reducing the large quantities of pulses and oilseeds imported every year.

With water and energy shortages fast mounting, it is imperative to cut wastages and improve their use efficiency. Holistic farming optimally utilizes available moisture, and vastly improves national water and energy efficiency.

Scientists estimate that for every 1% of organic matter content, the soil can hold 16,500 gallons of 'plant available water' per acre of soil of one foot depth (Source ATTRA) Groundwater recharge is also enhanced by increased biomass and mulch, correspondingly diminishing run-offs and floods.

## 11 POINT AGENDA FOR THE 11TH PLAN

Considering the above compelling reasons for adopting a holistic approach favoring sustainability, national health and distributive justice, we strongly recommend and petition the Government of India to adopt an 11 point programme on agriculture for the 11th Plan as follows:

(1) Holistic sustainable farming should be promoted vigorously to cover at least 20% of all farmland progressively every year, so that the major part of the cultivable area of India is converted to holistic, ecologically and economically sound and sustainably productive farming systems by the end of the 11th Plan. Government policies and programmes should be specifically prioritized, designed and targeted to help the small and marginal farmers, who constitute almost 80% of farmers, to achieve food and livelihood security. Even small holdings have demonstrated that they can efficiently meet food and livelihood needs with proper policy and support.

(2) The current excessive use of inorganic nitrogenous fertilizer that is incrementally destroying the vital biological quality of Indian soils, and contributing to global warming, should be discouraged by phasing out the subsidy on urea through a progressive 20-25% reduction each year, publicly announced in advance. Similarly, subsidies for all other agro-chemicals and heavy farm machinery should also be phased out within 5 years.

(3) All the money saved by reducing/withdrawing subsidies as aforesaid (Rs 22000 crore per annum for urea alone) should be used as incentive/support for farmers who adopt holistic organic farming This should include subsidy for local generation of bio-fertilisers, including the planting of trees for enhanced availability of biomass and nutrients. Each farmer adopting holistic, organic agriculture, helps build up the nation's agricultural capital of soil and water, but faces some risks and costs in the initial 2 to 3 years of transition period commonly needed for ecological and economic recovery. Soil-depleting chemical fertilizers have been subsidized all these years, and it is therefore warranted that a similar subsidy should be extended to those who produce their own fertilizers and enhance soil capacity for sustainable productivity. A cash

incentive to holistic farmers of Rs. 4,000 per ha per year for 2 or 3 years is strongly recommended to encourage this vitally necessary shift.

(4) A country-wide campaign must be undertaken on a war footing to regenerate our natural resources of soil, water, biodiversity, tree/forest cover, etc. Natural forests, grasslands and wetlands must be conserved and regenerated as they shelter fauna, avifauna and insects vital for agriculture, besides creating new fertile soil and augmenting recharge of groundwater. Outstanding benefits can thereby be efficiently achieved in the shortest possible time with significant savings in expenditure.

(5) Decentralised water harvesting and its optimal and equitable use must be actively promoted and financed by Govt policies, programmes and fiscal incentives/disincentives. Since at least 60% of Indian farming is rain-fed, the small and marginal farmer is at the mercy of an increasingly erratic monsoon. The first priority should be protective irrigation, through full subsidy for small farm ponds and/or ground water recharge by farmers This will reduce monsoon crop failure and facilitate introduction of a winter crop, thereby greatly reducing risks and increasing agricultural production and farmer incomes. This can be financed by decreasing the proposed massive and wasteful expenditures on large irrigation projects, having long gestation periods, high inefficiencies, fiscal irregularities and high social displacement, rehabilitation and ecological costs. Farmers need immediate provision of water on-farm, not through interlinking of rivers. Where irrigation projects exist, water delivery must be regulated through suitable local bodies like Water Users Societies, SHGs (Self Help Groups) and/or Rural Producer Companies controlled by locally resident farmers. A differential, incremental pricing mechanism should be mandatory for higher per capita /per acre levels of consumption of surface or ground water in excess of the minimum/basic allotment. (The water used to irrigate one acre of sugarcane is sufficient for at least 25 acres of jowar, bajra or maize.)

(6) The main objective of EGS, grants, subsidies, loans and Govt schemes should be asset creation for the small and marginal farmers. Delivery should be channeled through local SHGs, associations, or bodies of farmers. These, as well as Producer Companies that are wholly owned and controlled by locally resident farmers and which can engage the services of professionals to handle marketing, finance and other support services, must be actively promoted and supported. Such support/ incentives to local farmers' bodies should be proportionate to the number of farmer members represented by each such recipient body,

with additional benefits to those consisting exclusively of small and marginal farmers primarily engaged in natural resource regenerating holistic agriculture.

(7) The financial capacity of small Indian farmers, who are at the mercy of the market, should be strengthened so that they can hold their surplus produce till the prices are remunerative, and be helped to take up activities to add value to their produce through basic processing, packing, and collective marketing. Some Cooperative Banks, which provide grain storage and loans to farmers against grain stocks, are examples to be up-scaled. Government should promote the formation of local bodies/ Producer Companies of farmers and help them meet their needs for working capital, storage structures, community banks for grain, seed, fodder and other biomass, basic food and biomass processing units, tree planting, rainwater harvesting, biodiversity regeneration, extension and education, subsidy on small machinery/ farming tools/aids, oxen, bullock carts, etc. The Government has long provided generous incentives for the promotion of industry. Bodies of small and marginal farmers engaged in holistic farming should similarly receive tax waivers/concessions, low interest loans, and easier access to credit. There should be a farmer-friendly, single window delivery system for credit and Govt schemes through the rural banking network, which must be rapidly extended to support farmer groups. Organic Certification must be provided free. In fact, certification of chemically cultivated and GM crops, and cautionary labeling of those that are packaged, should be mandatory in the interests of consumers.

(8) Low cost and nutritious food must become available to all, and particularly to the poor. The Public Distribution System should offer a wide variety of locally produced nutritious crops and be managed by local SHGs or Producer Companies under civil regulation. This will reduce the high fuel cost and carbon dioxide generation of long-distance transportation. There should be total liberalization for primary processing, free movement and regional marketing of farm produce within a 150-200 km radius from farm to retail outlet, and proactive government support for basic transportation and regional marketing infrastructure which shortens the supply chain between farmers and consumers.

(9) Good Agricultural Practices used by innovative holistic farmers should be widely, systematically and urgently disseminated by use of Govt institutions, media, personnel and funds. Since the knowledge base for holistic farming systems is to be found among experienced, innovative

farmers, a partnership should be forged between them and existent educational, research and extension facilities to document, use and disseminate this knowledge. This can be accomplished faster by granting/leasing 10% of the land of Agricultural Universities and KVKs to innovative farmers with at least 3 years of demonstrated success in holistic farming for demonstration and/or extension of low cost holistic farming systems, including agriculture, horticulture, livestock and rainwater management, etc. Agricultural scientists and extension workers should help in scientifically documenting and widely disseminating this information.

(10) Genetically modified organisms/seeds must be banned in India – or, at the very least, strongly discouraged through the most stringent regulation and heavy, deterrent penalties for any spread of genetic pollution. They are likely to create new and more damaging cycles of ecological and economic risks. A handful of multi-nationals and Indian corporates are inducing farmers to adopt this high-cost, high-risk, poorly regulated technology by offering credit and promises of short-term benefits. Consumers across the world have been rejecting GM food and increasingly demanding poison free organic food. This is India's opportunity to be a leading producer of organic food and avail of this fast growing domestic and global market. The promotion of GM and GRT destroys this capacity and must be immediately halted and then reversed.

(11) Policies which permit the dumping of heavily subsidized agricultural products in India by developed countries – facilitated by the WTO – or which promote cash crop monocultures for export, making Indian farmers vulnerable to the disastrous consequences of any fall in global prices, must be discontinued.

Promoting holistic farming systems - which can provide food security to the 650 million Indians dependent on agriculture and nutritious food to all consumers- can effectively counter this threat to the nation's food security and farmers' livelihood and survival. It should therefore be the keystone of our agricultural policy, practices and priorities.

Holistic farming involves knowledge, native wisdom, and labour, rather than external industrial inputs or large per capita finance - with their attendant high costs and risks of indebtedness.

It can create significantly higher employment opportunities in the rural sector, thereby halting and reversing migration from rural to urban areas. It benefits the small and marginal farmers, who are the " Aam Admi" of the farming community. Government funds and resources should therefore be made available to them in proportion to their percentage in the rural population to honour the promises made in the Common Minimum Programme of the UPA government. Expropriating their land, the only asset of small and marginal farmers, by the Government for SEZs, and providing huge fiscal incentives to their promoters, is a policy that would inevitably lead to further pauperization of the farming community and the growth of armed insurgence.

Empowering farmers to ensure their own food and livelihood security through holistic farming systems, and through dispersed small industry based on agricultural produce, is the real need of the hour and the demand of the farmers, who constitute 65% of India's population.

All over the world, the demand for toxin-free produce is rising at an exponential rate of growth. Consumers in India are also entitled to it. Many States, and the farmers themselves, would be keen to adopt holistic farming systems if necessary support, as stated above, is made available. The Planning Commission, the Government of India, and the leaders of all political parties will fail in their duty if appropriate actions are not taken urgently to halt the promotion of GRT and GM agriculture, and to introduce the low-cost, low-risk, holistic farming systems which can save the small farmers from desperation and suicide. We thus earnestly call for your support in implementing the above proposals in the interests of the farming community and the nation as a whole.

## MAKING INDIA AN ECONOMIC SUPER POWER

For a nation like India, politically subjugated for centuries by alien conquest, and socially unsettled by invasions, the idea of becoming a super power is no ordinary thought. Understandably, even after winning freedom, India never aspired seriously to become a super power, politically or economically. Such an idea, if it existed, was dormant in the Indian consciousness. But it did acquire seriousness after the Pokhran blasts in May, 1998.

But even now, the national opinion is not uniform on how to become a super power. "Not atom bombs, only development can make India a super power," say some. "We need both to make us a super power", say others. Nevertheless, the idea that we should aim at being a super power is no more a dormant one, even if it is not a dominant one. But the desire has come a little late. There is background to it — the background of India's post-Independence history.

The freedom movement had promised much more than a mere change of rulers. But what happened was, as aptly described by English historians, a mere "transfer of power" from the aliens to the locals. Nevertheless, what did India intend to be after it became free?

When India attained freedom, it thought of emerging as a global leader, without becoming a global power. Its claim to leadership rested on the age-old Indian values of universal tolerance, peace and happiness. But the post-War world afflicted by cold war had no reverence for such high values. So India was swiftly marginalised in a world which respected only power.

But, within India, the Indian leadership did the other way round — it persuaded the people not to pursue their age-old values, but, accept the Anglo-Saxon ideas and institutions in the main. It folded back the philosophic lead shown by Gandhi, Aurobindo and Tilak. Their definition of the Indian identity was substituted by the western ideas of secularism and socialism. Since then, for over four decades, the Left-Socialist parties and intellectuals mounted a vicious attack on the Indian past, and virtually delinked the Indian polity, economy, history and education from its past and turned to Anglo-Saxon values.

This is precisely what the Indian freedom movement had struggled against. Because of this drift, to the Indian intellectual, India's past became a burden and ceased to be matter of pride. Also, the secular-socialist leadership systematically fragmented the Indian society into majority and minority, rich and poor, forwards and backwards, and denied India the deeper awareness of its intrinsic unity brought about by Hindu values and civilisation. This resulted in setting one Indian against another, leading to massive self-deprecation. This eroded our self-confidence as a nation. The idea of a powerful India could not be internalised in a situation where every Indian was running down the other.

The socialist regime turned two generations of potential entrepreneurs into job-seekers. Permit quota and licence, not efficiency or merit, became the route to business success. The competitive strength of India was systematically weakened, and the traditional skills of the trading and business communities in India were dissipated. The devastation caused by the socialist regime in the post-Independence period was more pervasive than the devastation of the Indian economy by the British rule.

The pre-British India, described as caste-ridden, feudalistic and unmodern, was economically ahead of the rest of the world,including Britain and the USA. The Indian economy had a share of 19 per cent of global production in 1830, and 18 per cent of global trade, when the share of Britain was 8 per cent in production and 9 per cent in trade, and that of USA, 2 per cent in production and 1 per cent in trade. India had hundreds of thousands of village schools and had a functioning literacy rate of over 30 per cent. In contrast, when the

British left, India's share of world production and trade declined to less than 1 per cent and its literacy was down to 17 per cent. And yet, in 1947, India had large Sterling reserves, no foreign debt, and Indians still had an effective presence in such trade centres as Singapore, Hong Kong, Penang, Rangoon and Colombo. But, by the time the Socialist regime came to a close, India had become politically and economically weak and disoriented, lacking in self-confidence. Its influence in South Asia too had waned.

But, with the collapse of the Soviet Union, the Indian establishment swung to the other extreme — from socialism to market capitalism. It was again in tune with its lack of self-confidence — if it is not communism, it has to be capitalism. And we began to witness a massive effort to globalise — which means westernise — the Indian economy. The persons and parties who grossly maladministered the national economy suddenly turned 'reformers' and they paraded themselves as the new messiahs as if someone else was responsible for the wrongdoings in the past which needed to be reformed. Thus, Congressmen, that is those who needed to be reformed, styled themselves as 'the reformers'. Just as socialism was an ideology earlier, globalisation too became an ideology. Now anyone who dissents against globalisation is being labelled as backward-looking and anti-modern. Thus, just as we adopted socialism without a debate as to its viability in India and discarded it, we have adopted globalisation without any debate. As a result, we have set out blindly to experiment on India the experience of the West. And just as we had committed one set of mistakes before 1991, we have now begun committing another set of mistakes from 1991.

The great question remains: Will what we have been doing since 1991, make India an economic super power? If not, by what method will India emerge as an economic super power? Before we comprehend what India should do to become a global power, we must understand properly what is happening in the world today.

There is a gradual shift in the global view on global trade and currency regimes of today. It questions not just the global assumptions, but also the national assumptions about the global regime. The experience of Mexico first, the EC currency crisis next, and the South Asian crisis now, tend to question the very foundation, stability and viability of the present global economic regime under installation. The assumptions about what India should do is also changing fast. For instance, not even the diehard lobbyist for globalisation today talks about the full convertibility of Indian currency for which P. Chidambaram had, only a year ago, unveiled a three-year master plan. Dr Jagdish Bhagwati, an uncompromising advocate of currency convertibility, has now come to the view that currency convertibility should be "avoided like a plague". This single shift is enough to demonstrate how unstable are global economic prescriptions and institutions.

It is not that what happened in South Asia now was not anticipated. Akio Morita, the former Chairman of Sony Corporation, wrote a letter to the G7 leaders in 1992 questioning the rationale of globalisation with a currency regime operated by mindless speculation. Akio Morita said that what manufacturing, efficiency and technology achieved might be destroyed by the free-for-all money market over which no one had any control. The world was forced into the present free-for-all currency regime in 1971. It was to prevent this, and to prevent money turning as a destabiliser that the world opted for a fixed rate of exchange in 1949. The very constitution of the IMF, the World Bank and GATT after the World War II was to avoid a free-for-all currency regime. But it was reborn in the year 1971 after the gold-dollar convertibility was discontinued. But, with global economics coming under money-spinners, the new regime has been increasingly questioned, more seriously now.

What is the lesson for India in the changing situation? The WTO, which is the symbol of free trade, cannot function without a free currency market, but, a free currency market destabilises free trade. This is a Catch-22 situation for which there is no answer within the global trade regime today. A mere comparison between the trade volumes in the world currency market and the volumes of trade in goods and services is adequate to demonstrate how paper money dominates trade. The daily global trade in currency is over $ 1.2 trillion; in contrast, the annual global trade in goods and services is $ 4.5 trillions. It means that for every dollar of actual trade, the trade and speculation in currency is $ 95!

If anything that will ultimately destroy the global trade regime, it will be the global currency regime. That means the present global trade regime does not have a guaranteed success, though it is not guaranteed to fail. India should not reorganise its national economy to match a global regime which has no guarantee for success. It means that India shall have to frame a national agenda.

So, the road to economic success in future will be not unlimited globalisation, but, controlled, guided, calibrated globalisation which is the vehicle for acquiring global economic strength. Even in the past, it has been so. Japan became an economic power by controlled globalisation even as China is doing the same thing now. It means that India has to devise its own strategy, that is what the idea of Swadeshi is all about. Swadeshi means Indian agenda to which the global agenda can only be a supplement, and not the other way round. This is what China did in the 1980s, it set out on a national agenda to double its food production from 200 million tonnes to 400 million tonnes between 1983 and 1993. It turned to its agriculture to power its economic growth beyond 10 per cent during the 1980s and turned to other areas to sustain the growth rate afterwards.

In contrast, the Indian agriculture virtually remained stagnant throughout the period from 1990 to 1998. This is how we malprioritised our economic strategy even after we set out to desocialise our economy in 1991. What then should we do now?

The answer is obvious. If we have to break into a high trajectory growth rate on a sustained basis, just as China did, we have to start with agriculture. Higher agricultural growth leads to rural prosperity which, in turn, leads to more buying power, and that triggers high industrial growth. The Indian agriculture has greater advantages than the Chinese agriculture had. We have one-and-a-half-times as much plain land as China has and a whole year's sunshine. Agriculture involves the greatest number in our country and unless a nation activates a majority of its people, it cannot grow. Thus, agriculture is our biggest asset. We should have invested more and more capital in agriculture. But, in contrast, the capital formation in Indian agriculture is declining from year to year down from 17 per cent in 1980 to 9 per cent now. So agriculture in India has been increasingly neglected. Agriculture is the starting point for India to emerge as an economic super power.

The second trigger for high growth is the energy sector. It is linked to transport. We did not plan for multimodel transport, nor did we integrate energy and transport in planning. We have allowed the cheaper Railways to decline in favour of costly road transport. The Railways share in freight traffic has come down from 88 per cent in 1950 to 37 per cent now; in passenger traffic from 68 per cent to 13 per cent. The reduction in the Railways share has meant a corresponding rise in the share of energy — intensive road transport. Why did the Railways decline? It is because we have saddled our Railways, which is one-and-a-half-times in length of the US Railways and twice as that of China, with coal transport and coal accounts for 45 per cent of railway freight today. It completely clogs the railway system and prevents its modernisation. We have not yet tapped the vast potential of coastal transport and inland water transport which can considerably ease the pressure on the energy-intensive road transport and the clogged railways.

Water transport involves less than 1/3 of the energy as land transport. We have not also tapped atomic energy. The expert view is that by using fast-breeder reactors with international safety standards, India can produce 3,00,000 MW of atomic power with indigenous resources. Our current power generation from all sources is about 89,000 MW. Atomic energy has higher initial investment, but, no running cost. But in thermal power, running cost is as high as 50 per cent of the production cost. In other countries, atomic energy constitutes a high percentage of total energy — it is 76 per cent in France, 40 per cent in Sweden, 29 per cent in Japan and 23 per cent in USA, but less than 2 per cent in India. Also we can turn alcohol into diesel like Brazil does, in large quantities. An integrated energy-cum-transport planning can trigger a high growth.

The third area for large-scale activation of India is the non-corporate sector which accounts for 40 per cent of our GDP, over 30 per cent of exports and more than 75 per cent of employment outside agriculture. This sector consists of the traditional craftsmen and artisans in the main. During 1997-98, this sector had contributed over $12 billion by way of remittance from skilled Indian workers employed abroad. From 1980, the total remittance by this sector has exceeded $ 75 billion. This is largely self-employed and labour intensive. This has withstood the socialist assault and, later, the free market globalisation.Yet this sector has had no assistance from the system by way of organised finance or employment or social insurance schemes. This has enormous potentiality for high growth.

As an adjunct to this is the services sector — both modern and traditional services. This is an area of high growth potentiality for India because of the advantage of familiarity with the English language, particularly in the modern services sector like law, accounting, consultancy and advertising. And yet we are mindlessly allowing the MNCs to enter this area where we have our dominance.

Yet another area in which India is expected to become a major player is textiles, particularly after the quantitative restrictions are removed by the West from 2005. Textiles account for 30 per cent of India's exports and the largest industrial employment. But this crucial area is underdeveloped in terms of technology and processing. Powerlooms, which account for 73 per cent of the textile produced in India, are not funded by organised financing. Organised finance, technical upgradation and freedom from tax inspectors will transform the textile sector into a high growth area.

Computer software and IT technology are an important area. The Indians in the Silicon Valley in the USA have proved our proficiency in IT products. The global players are sourcing India in a big way because of the highly skilled manpower available here. Some of the best minds in India are taking to this area of expertise. Since information is key to the world ahead, the Indian mastery over this area is likely to make India a dominant IT power in the next decade.

There are estimates that by 2002, India will export software products worth $ 10 billion.

This takes us to the role of foreign direct investments in India emerging as a global power. By its very nature, FDI has a marginal, but, crucial role to play. But it must be made to flow into areas where the local capital is shy or incapable of funding a mega venture, particularly for global use. It can usefully supplement our large national savings which, in dollar terms, amount to over 90 billion. Instead of going for FDI in areas which hurt the local economy and employment, the government may allow 100 per cent FDI in, say, four large international airports or two large seaports. Such facilities require income to be generated from abroad for their very survival. Indian capital cannot build them and the ordinary Indian need not pay for them. Such airports alone can fetch an FDI of 20-30 billion dollars in the next five years. The thrust of the FDI regime must be to channel FDI where local capital is inadequate and from where the FDIs cannot leave India at will.

Indians must realise, as have the westerners and the Japanese, that global business is war, not just commerce. Wars have always followed commerce and commerce has followed wars. Therefore, trade, like war, is a game of deception; and free trade is a strategy, not an ideology. That is why nations build up non-tariff barriers. South Korea used to tax-raid the buyers of foreign cars. The Philippines mandates that cars may be produced by foreign companies, but, distribution can only be by a local company.

France has instituted professional appraisal of imported electronic goods so as to clear only a fraction of the imports. Thus, every nation digs trenches to fight the advances of global trade to the prejudice of national interests. But in India, just as socialism did, free trade has emerged as an ideology. Indians must learn to look upon trade as war and free trade as a strategy and develop native non-tariff formulations.

Again, what India and Indians critically need is a measure of national pride and self-confidence. Without both, no nation can prosper. It is unfortunate that the English-educated Indians have repeatedly failed to generate personal or national self-confidence. Instead, they began to deprecate everything Indian. In the process, they deprecated each other, resulting in all-round self-deprecation. This trend has to be reversed. If we deprecate ourselves, as we have been habituated to, no one is going to respect India. The atomic test by India triggered our national pride in a big way, but we threw cold water on it. We have to build, and not waste, such achievements whether it is the Param Super Computer, or PSLV launch, the Tata Car or Kamini nuclear reactor working on thorium. A rapid build-up of national pride and self-confidence will yield a quantum jump in economic advancement. Trade is a matter of psychology.

When Japan followed the Chinese lead in technology, it prided itself as combining the 'Japanese spirit' and 'Chinese technology'; when it was forced to accept the western technology, it coined a new motto, 'Japanese spirit with western technology'. Thus, a measure of nationalism has to be built up to create a psychology of national pride. This is needed for economic advancement and those who delink national pride from economic growth will have to be educated on what the world history teaches every nation.

And, finally, what India needs, as the first step, to transform into a global economic power is national consensus. The confrontationist politics and political untouchability that we are witnessing for the past few decades has made the formation of national consensus difficult. But this is what the Japanese considered as the first step for national economic development. In fact, even the Japanese democracy is unlike the Anglo-Saxon model. It refused to accept the divisive idea of majority-minority party politics and continues to pursue, even today, consensus as the national principle.

But the Indian parliamentary democracy, which is a carbon copy of the western system, does not allow the emergence of national consensus in politics; even though the Indian society functions on consensus, the Indian polity does not. For transforming India into a global power expeditiously India has to review its polity and bring it in tune with its traditional principle of panchayat or the idea of consensus to which the Indian society is accustomed.

There is no need for a trade-off between democracy and development, as Lee Kwan Yew, the former President of Singapore, proposed. Probably, a presidential form of government will ensure the support that national economics needs from national politics. Yes, India has all the potentialities and also the compulsion to become a global economic power in the 21st century, and early enough in the coming millennium.

All that India expects of its leadership is that it should ensure that there are no round pegs in square holes. The leadership must inspire the nation to think and function as a nation. The essential quality of a leader is that he does not care how long will he be in power, but how decisively he rules. This is what Lal Bahadur Shastri did. He ruled for just 18 months and asked the people to starve on Mondays — a trans-religious auspicious day. He is remembered for both. India is expecting such a trait in its leadership to lay its claims to becoming an economic super power. Will the present leadership, which claims to be different from the previous ones, respond to the will of the nation to become a global economic power?

## SUSTAINABLE DEVELOPMENT OF RAINFED AGRICULTURE

India's agricultural growth has been sufficient to move the country from severe food crises of the 1960s to aggregate food surpluses today. Underlying this growth were massive public investments in irrigation, agricultural research and extension, rural infrastructure, farm credit and rural development programs. India's agricultural sector, however, faces severe challenges for the future. Despite sizeable national food stocks (30 million tons in 1995), widespread poverty and hunger remain because agricultural and national economic growth have not adequately benefitted disadvantaged regions and the poor.

The demand for basic staples, non-food grains, and exports is increasing. At the same time, resources are shrinking and the productivity of some resources already being utilized is threatened by environmental degradation. Growth in total factor productivity is reported to have declined slightly in major crops. Returns to investment in agricultural research and rural infrastructure are reported to be high, but these investments remain low.

Most of the increase in agricultural output over the years has taken place under irrigated conditions. The opportunities for continued expansion of irrigated area are limited, however, so Indian planners increasingly are looking to rainfed, or unirrigated agriculture to help meet the rising demand for food projected over the next several decades. Despite the historic bias in favor of irrigated agriculture in terms of research and infrastructural investments, rainfed agriculture has always been an important part of the agricultural sector.

Rainfed agriculture accounts for about two-thirds of total cropped area (Government of India 1994b, nearly half of the total value of agricultural output. Nearly half of all food grains are grown under rainfed conditions, and hundreds of millions of poor rural people depend on rainfed agriculture as the primary source of their livelihoods.

Rainfed areas are highly diverse, ranging from resource-rich areas with good agricultural potential to resource-poor areas with much more restricted potential. Some resource-rich rainfed areas potentially are highly productive and already have experienced widespread adoption of improved seeds. In drier, less favorable areas, on the other hand, productivity growth has lagged behind, and there is widespread poverty and degradation of natural resources. Even given that rainfed agriculture should receive greater emphasis in public investments, a key issue is how much investment should be allocated among different types of rainfed agriculture. Outmigration and income diversification into the nonagricultural sector must provide the long term solution to economic development of many resource poor areas, but these opportunities currently are inadequate in relation to population growth to provide short to medium term solutions. Agricultural growth in these areas will be essential for reducing poverty and environmental problems in the decades ahead.

There is a need to identify the opportunities for stimulating agricultural growth and reducing poverty and environmental degradation in rainfed areas. Likewise, there is a need to assess the opportunity costs of diverting scarce public resources from resource-rich to resource-poor areas.

The tradeoffs between investing in resource-rich and resource-poor areas in terms of their productivity, poverty and environmental outcomes need to be understood in order to guide public policy decisions toward productive outcomes.

Developing strategies for rainfed areas is difficult because of their diversity in terms of agroecological characteristics, infrastructural development, and other socioeconomic variables. On an all-India scale, for example, rainfed systems include high-rainfall agriculture in the east and northeast as well as the drought-prone areas of the Deccan Plateau. Other agroclimatic characteristics such as soil types also vary, as do infrastructure development, human capital, and other socioeconomic factors.

Across villages within a district, for example, there is wide variation in access to paved roads and public transportation to market centers. Similar diversity of agricultural systems is found even at the local level. Individual villages in the semi-arid regions, for example, often contain numerous soil types with widely differing crop production potential (Dvorak 1988). Irrigation wells are found in practically every village, so irrigated and rainfed agriculture co-exist almost everywhere.

Diversity at both the national and local scale has implications for agricultural development strategies. At the national scale, there is a need to distinguish among regions according to their constraints to agricultural development. This requires creating a typology of rainfed agriculture that would incorporate both agroecological and socioeconomic variables in order to serve as a tool for planning agricultural research and other public investments. Local-level diversity of rainfed agricultural systems, meanwhile, implies that planners must recognize that changes in policy, technology or infrastructure may have varying impacts across small areas, and that there is a limit to the extent to which external interventions can induce finely-tuned responses. Regional or district-level planning must be complemented by local initiatives that can be more responsive to specific needs.

## OBJECTIVES OF THE PAPER

The overall goal of this paper is to review the important issues in rainfed agricultural development and report on the progress made in India to date. This will serve as a precursor to a detailed study to be carried out by the Indian Council of Agricultural Research, IFPRI, ICRISAT and the World Bank. That study will result in recommendations for designing a strategy to develop rainfed agriculture in India. In this paper, we compare the past performance of rainfed and irrigated agriculture and of different types of rainfed agriculture, including relatively high- and low-potential areas. We attempt to identify the factors that determine differences in performance, and we examine the possibilities for influencing those factors.

Where information is not available, we recommend further analysis that may be required as a prerequisite to formulating a thorough strategy for rainfed agricultural development. We approach the problem by reviewing the relevant literature on the subject and conducting a statistical analysis of all-India district-level data.

The database contains several agroclimatic and socioeconomic variables that we hypothesize to influence agricultural performance at the district level. The district-level approach, of course, does not permit analysis of the implications of micro-level diversity of rainfed agricultural systems.

Therefore we focus on broader indicators with implications for area-wide development efforts. In addition to the district level data analysis, we review the literature on rainfed agricultural performance in terms of technology adoption and performance, yield levels and their variability, natural resource sustainability, and poverty alleviation. We examine the role of economic and social policies, area development programs and infrastructural investments in promoting sustainable rainfed agricultural development. On some topics sufficient evidence is available to draw conclusions and make policy recommendations, and on others, additional analysis is recommended.

## CHARACTERISTICS OF RAINFED AGRICULTURE

In this section we introduce some characteristics of rainfed agriculture that will influence our approach to the problem of deriving recommendations to stimulate rainfed agricultural growth.

As mentioned in the previous section, rainfed and irrigated agriculture coexist in practically every village in India. Public investment programs, however, usually cannot be targeted so precisely. For practical purposes, they need to be planned and implemented on a larger scale, such as at the village, taluk, district or state level. For example, public programs that provide credit, employ people, or build roads cannot target their efforts to either rainfed of irrigated agriculture; they can only target areas that are relatively more irrigated or more rainfed. Price policies, on the other hand, can attempt to target rainfed or irrigated agriculture within a given location by targeting crops that may be more likely to be rainfed or irrigated. But few crops are either 100% irrigated or 100% dryland, so some spillover will always remain. Also, every crop is grown over a large geographic area, so it is difficult to isolate the socioeconomic and agroclimatic variables affecting their performance. And while price policies are important, they are not the only approach through which policy makers can influence agricultural development.

## DEFINING RAINFED AGRICULTURE

In the present investigation we use the district as the unit of analysis. We do so for two main reasons. First, through a district level focus our analysis may be relevant for public investment programs that provide infrastructure or other social services to particular areas. Second, a district focus enables us to examine the contributions of both socioeconomic and agroclimatic variables on a nation-wide basis.

The district is the smallest administrative unit for which the required data are available. To arrive at a district-level definition of rainfed agriculture, we consider the percentage of each district that is irrigated or rainfed, and consider predominantly rainfed districts as "rainfed" and predominantly irrigated districts as "irrigated." Obviously there is a certain degree of arbitrariness to any threshold we may choose to distinguish between irrigated and rainfed districts. Several previous studies have faced this same dilemma in categorizing rainfed areas. Some of them have distinguished between irrigated and rainfed districts according to certain criteria such as the amount of rainfall and the level of irrigation.

As mentioned above, all of these definitions suffer from the inability to distinguish between rainfed and irrigated agriculture within districts, but we accept this as an inevitable limitation. Another problem is that both the rainfall and irrigation thresholds are defined somewhat arbitrarily. We discuss rainfall and irrigation thresholds in turn.

### Rainfall Criteria

At the same time, the 500-1500 mm range maintains a degree of homogeneity in the types of agriculture under analysis by excluding both very dry, desert areas and very high rainfall areas. Such areas may face unique constraints that limit comparability to agriculture under the more moderate conditions that predominate in most of the country.

Shah and Sah (1993) and Thorat (1993) use narrow rainfall thresholds with a relatively low maximum because they intended to focus on very dry (but not quite desert) areas. It is important to note that the impact of the level of rainfall on crop production is conditioned by both the distribution of rainfall over the course of the season and the factors that determine moisture retention in a given location. As a result, narrow rainfall thresholds such as those used by Thorat (1993) are likely to combine some areas with disparate moisture regimes and separate others with similar moisture regimes.

A narrow rainfall range probably makes sense only if limited to relatively uniform soil types. Incorporating soil types into the definition, however, introduces yet another variable and makes the definition somewhat clumsy.1 For this reason, our preference is to utilize a relatively broad range of rainfall levels such as used by Bapna et al, (1984), excluding only those that are either desert environments or extremely humid. We choose a range of 450-1600 mm average annual rainfall as the range for predominantly rainfed districts. The lower bound of 450 mm excludes the desert districts of western Rajasthan as well as one district each in Punjab, Haryana and Gujarat. The 1600 mm upper bound excludes the Himalayas, the northeastern states, Kerala, all the coastal districts of Karnataka and Maharashtra, and one coastal district each in Tamil Nadu and Gujarat. Where more disaggregated analysis is required to examine the performance of relatively moist or dry rainfed areas, we can subdivide the rainfall criteria into a low rainfall area (<750 mm per annum), a medium rainfall area (750-1125 mm), and a high rainfall area (>1125 mm), sometimes described as the arid, semi-arid and humid areas. To repeat the earlier caveat, these broad rainfall classes are heavily conditioned by the factors that determine moisture retention, particularly soil type.

## Irrigated Area Criteria

Classifying districts by irrigated area is difficult for several reasons. First, any threshold percentage area irrigated must be defined somewhat arbitrarily, and second, in most districts irrigated area has increased steadily during the period under study.

In these studies the threshold ranges from 10 percent to 30 percent, with most defining rainfed districts as those with less than 25 percent irrigated area. 25 percent is the mean irrigated area for the years 1956-90, so according to this definition, rainfed areas are those with less than average area irrigated, while irrigated areas are those with more than average area irrigated. A number of arguments can be made about whether the figure of 25 percent is appropriate, but ultimately any definition based on such a threshold suffers from the problem that slight differences in irrigation levels will move some districts from one category to the other.

One approach is to use three categories of irrigation instead of two in order to more clearly identify the characteristics of lightly and heavily irrigated districts. This approach however, also has weaknesses, because more categories means more thresholds, which means that fewer districts will remain in one category over the entire period. As a result, when the three irrigation categories are used as described above, nearly half of all districts shift categories at some point in the period. When only two irrigation categories are used, on the other hand, only about one quarter of the districts shift categories.

For that reason, we also conduct the analysis with only two irrigation categories, with districts with less than 25 percent area irrigated considered unirrigated, and districts with 25 percent or more area irrigated considered irrigated. We conduct the analysis twice; in one case all districts are analyzed, and in the other only districts that remain in one category or the other throughout the study period are used.

Before continuing, we briefly discuss two other criteria for defining rainfed areas that we considered but rejected. First, districts could also be subdivided by the extent of different types of irrigation, in particular, canals, wells, or tanks. The justification for this concerns the quality of irrigation services delivered to each farm. Well irrigation, for example, is controlled by the individual farmer (to the extent that the aquifer yields water), whereas under canal irrigation farmers depend more on the amount of water taken by their upstream neighbors, so they incur a greater risk of drought.

As a result, farmers with well irrigation apply more inputs and have much higher yields on average (Shah, 1993). The district-level data cover gross cropped area as opposed to net cropped area, so we control for variations in the quantity of irrigation water delivered to the farm. However, the data do not control for variations in quality or differences in farmers' response to the different levels of risk under each irrigation source. When the circumstances warrant it we may examine irrigation by source, but mainly we do not, since our main focus is on rainfed agriculture, not distinctions in irrigated types.

A second alternative criterion for defining irrigation levels would be to distinguish districts by the proportion of farmers who have access to some irrigation. In many areas, water markets or shared irrigation wells enable farmers to gain access to irrigation even if they do not own a well or are not directly serviced by a canal or a tank. As a result, the number of farmers with access to irrigation can be much larger than the number who own wells (Shah 1993).

The distinction between the proportion of area irrigated and the proportion of farmers is important because it can affect the way in which most farmers manage their crops. If a larger proportion of farmers have access to a small amount of irrigation, they may concentrate their managerial and other inputs on irrigated plots, which they may perceive to be more productive and less prone to risk of crop failure than dry plots.

In this case, perhaps dryland crops would be less productive (though more farmers will be better off). Farmers without access to irrigation, on the other hand, may devote relatively more resources to rainfed crops than those with some irrigation. Unfortunately we do not have access to district-level data on the proportion of farmers with access to irrigation, and we do not know if the proportion of farmers with access to irrigation and the proportion of area irrigated vary independently of each other across districts. Therefore we cannot consider using this definition; we raise it only to draw attention to some of the issues to consider in developing a definition of a rainfed district.

## THE NEED FOR A TYPOLOGY OF RAINFED AGRICULTURE

The definition of rainfed agriculture presented in the previous section will assist us in comparing the performance of predominantly rainfed and irrigated districts. We also wish to compare the performance within rainfed agricultural districts and relate differences to a range of constraints to agricultural development.

This in turn will be useful for prioritizing and organizing agricultural research, public investment, and policy and institutional reform.

To characterize districts according to the various agroclimatic and socioeconomic variables that constrain agricultural development, we will need to construct a typology based on those variables. In this section we discuss in a bit more detail the typologies of Indian rainfed agriculture that already exist, the reasons why a new typology needs to be constructed, and the ways in which such a typology would be used.

### Existing Typologies

Current typologies of Indian agriculture are based on agroecological zones. Recently, ICAR delineated 20 agroclimatic regions based on soils and climate (NBSS&LUP, 1992). 13 of these zones cover the area of this study; of the rest, 5 are in the Northeast, the Himalayas, and the Andaman, Nicobar and Lakshadweep Islands; one zone covers the high rainfall areas of the Western Ghats and the Arabian Sea coast; and one zone covers the desert in the western parts of Rajasthan and Gujarat.

ICAR's agroecological zoning system is based on variations in rainfall, soil type and temperature. Irrigation status, however, is conspicuously absent. This is a severe limitation due to the primary importance of irrigation in determining cropping patterns and productivity in most of the country. In this analysis we overcome that shortcoming by dividing zones into primarily rainfed and primarily irrigated districts. We discuss this division further below. The 20-zone system is sufficiently disaggregated to enable it to keep problems of within-zone variation to a manageable level, and the number of zones remains small enough to be manageable for most uses.

Also, the zones can be easily reaggregated for particular purposes. Recently ICAR subdivided the 20-zone typology into a total of about 50 subzones; such a disaggregated typology may be useful for certain agricultural research purposes, but for policy analysis it is too large to be functional.

In our analysis, using rainfed and irrigated districts in the 13 zones covered in our data would yield a total of 26 categories, which becomes unmanageable. For the purposes of our district-level analysis, we modify the 20-zone system so that it is sufficiently aggregated for our purposes. We create a new 5-zone system in which each zone is a combination of two or more of the zones of the 20-zone system.

The new zones, require some explanation. First, the 20-zone system does not follow district boundaries, but in our analysis districts must remain intact. As a result, in many cases we classify a district as lying in one zone even though part of it may actually lie in another. Districts in ICAR's Zone 18 along the Bay of Bengal coast, for example, also lie in adjoining agroclimatic zones 7,8 and 12.

As a result, zone 18 drops out of the sample. Second, by combining the zones it is inevitable that some new, aggregated zones will contain substantial within-zone diversity. We find that for the case of ICAR zone 4, it makes sense to place the portion of zone 4 that lies in the Gangetic plain in one new zone, and the part that lies in upland areas of Rajasthan and Gujarat in another.

## What a New Typology Should Look Like

The existing agroclimatic typologies may be adequate for a narrow set of objectives, such as locating where certain crops are likely to be produced and which regions may be prone to certain natural resource management problems.

Beyond such highly specific applications, however, the agroclimatic typologies are of limited use because they are so narrowly defined.

In order to be useful for designing a strategy to develop rainfed agriculture, a typology must be constructed on the basis of the whole range of factors that affect agricultural development. These extend far beyond simple agroclimatic conditions or even irrigation status. If a district has favorable growing conditions but lacks the infrastructure needed to support productive farming, for example, it should not be surprising to find poor performance in that district despite the favorable agroclimatic conditions. Later in the paper we will examine the determinants of agricultural performance and demonstrate that other factors in addition to agroclimatic conditions help explain the variation in performance. In addition to agroclimatic conditions and irrigation status, numerous additional variables can be hypothesized to influence agricultural development. Physical infrastructure such as roads and electrification, for example, and social infrastructure such as banks, markets and agricultural research and extension services, can be expected to play an important role in stimulating the agricultural sector. Demographic indicators such as population density and literacy levels also may be related to agricultural performance, as may economic policies that directly or indirectly affect input or output prices. Institutional considerations also may affect performance; they include laws governing trade, property rights, prices of inputs and outputs, etc., and the quality of services provided by government agencies.

Just as there are many determinants of performance of the agricultural sector, there also are many criteria for evaluating performance. Productivity growth is one that is commonly applied, but others include the levels of poverty and food security, the variability of production and income, and the degree of degradation of natural resources. The ideal typology of Indian agriculture would characterize regions or areas according to all the factors that determine performance over a broad range of criteria. In this way it could serve as a valuable planning tool for public investment in agricultural research, infrastructural development, poverty alleviation programs, policy and institutional reform, etc. While such a "super typology" might be unattainable, it presents an objective to work toward. In preparing this paper we lack the resources to develop an acceptable typology, but it remains a high priority for future research intended to support Indian rainfed agricultural development.

Later in the paper we analyze the determinants of rainfed agricultural development using multiple regression analysis. We will identify many of the agroclimatic and socioeconomic factors that contribute to performance of Indian rainfed agriculture according to a variety of criteria. We will stop short of creating a typology, but we will gain preliminary indications of the kinds of information that need to go into such a typology.

## INSTABILITY OF RAINFED VS. IRRIGATED AGRICULTURE

Agricultural yields are more unstable in rainfed areas than where irrigation is assured. But if irrigation water supplies fail, then irrigated agricultural area can fall sharply, leading to even more unstable output than under rainfed agriculture. Dhawan (1988a) used the coefficient of variation10 (CV) as the indicator of instability of production, yield and area cultivated for irrigated and rainfed food grains. Using national data, Dhawan found that the CV of output of unirrigated food grains for the period 1970-83 was 11.4%, significantly higher then the 6.4% for irrigated food grains. Decomposing these into output and yield instability, the CVs of yield were 9.2% and 5.3% for rainfed and irrigated food grains, respectively, and for area they were 3.1% and 2.3%, respectively. This suggests that yield and area moved together to increase output instability. Shah and Sah (1993) also found higher yield variations in rainfed than irrigated crops. They calculated CVs ranging from 36% to 70% for yields of rainfed food grain crops and 9%- 26% for irrigated food grains. They did not examine how these yield variations translated into production instability. The larger variations found by Shah and Sah most likely result from their small sample size of a few rainfed and a few irrigated districts over 10 years in the state of Gujarat. Dhawan's study was based on nationwide data, which is likely to be more smooth. Walker (1989b) found similarly large fluctuations in the yields of both rainfed and irrigated crops in a village level study in Maharashtra and Andhra Pradesh. This study was based on 40 households in each of three villages over 10 years, so it is not expected to be smooth. For irrigated paddy in one Andhra Pradesh village, the mean CV was 31% between 1975-76 and 1983-84, and for several rainfed crops in three villages in Andhra Pradesh and Maharashtra, the mean CVs ranged from 44% to 69%. In tank irrigated areas of south India, Hazell and Ramasamy (1991) found that paddy production fell 50% in their study villages in Tamil Nadu in the 1982-83 drought; virtually all of the decline resulted from reduced area planted. Bidinger et al (1990) found an even greater decline in paddy area in their study village in Andhra Pradesh in the 1985-87 drought, with little change in irrigated yields.

# 4

# Trends in Recent Agricultural Developments

## INTRODUCTION

Agriculture provides basic sustenance to all living beings. It is very important that ecologically, socially and economically sustainable agriculture should become the backbone of the development process of the State. Agriculture should be sustainable so that the natural resources such as soil, water and biodiversity are used efficiently and equitably. It should be economically viable and lead to increasing employment opportunity, socially feasible, strengthening the role of women and other marginalized sections of the people. Equity in sharing benefits is vital for community participation in the conservation and enhancement of natural resources.

The Government is aiming to achieve 100% food security in the State and also to create avenue for export of agricultural produce for economic upliftment of the farming community. During the Tenth Plan period, the state is aiming an annual growth rate of 4% in Agriculture and 8% in Horticulture crops for sustainable Agricultural development, employment generation and poverty alleviation.

The Government is focusing its policies towards overall development of agriculture sector in terms of increasing the cropping intensity by bringing every piece of land under cultivation, productivity increase, maximizing natural resources with parallel efforts to conserve them. The state aims at bring second green revolution in dry land Agriculture, while sustaining the tempo of agricultural development in irrigated agriculture. The change in land use pattern created serious concern among the agricultural planners to evolve suitable development strategies.

The agricultural performance of the state is given in the Table below. (Kg per ha.)

| Crop | Average Yield | | Highest yield in India | Tamil Nadu's place |
|---|---|---|---|---|
| | India | Tamil Nadu | | |
| Rice | 1804 | 3350 | 3510 (Punjab) | Second |
| Jowar (Cholam) | 769 | 962 | 962 (Tamil Nadu) | First |
| Bajra (Cumbu) | 610 | 1348 | 1348 (Tamil Nadu) | First |
| Redgram | 616 | 710 | 1301 (Bihar) | Fourth |
| Total Food-grains | 1562 | 2238 | 3830 (Punjab) | Fourth |
| Total Oilseeds | 710 | 1611 | 1611 (Tamil Nadu) | First |
| Groundnut | 733 | 1784 | 1784 (Tamil Nadu) | First |
| Cotton | 193 | 305 | 410 (Punjab) | Third |
| Sugarcane | 64562 | 106778 | 106778 (Tamil Nadu) | First |

Source: Agricultural Statistics at a glance 2004 - Agrl. Statistics Division - GOI

The increasing trend of fallow lands (both current and other fallows) which was 22.93 L.Ha. during 1990-91 to 24.35 L.Ha. during 2001-02 which was further increased to 29.93 L.Ha. and 28.17 L.a. during 2002-03 and 2003-04 respectively due to drought situation caused reduction in cropping intensity from the average level of 120% to 113.4% during last 2 years. The gross cropped area declined to 53.16 L.Ha. during 2003-04, from the average normal coverage of 63 L.ha. leaving about 9.8 L.Ha. under fallow. Though the reduction of cropped area during 2003-04 is evidently more due to acute drought prevailed during that year, still there is reduction under gross cropped area to the tune of around 4 L.Ha. between 1990-91 to 2001-02. Inspite of various constraints such as fragmentation of land holdings, more than 52% of the cultivable area are under dryland condition, over exploitation of irrigation sources, deterioration of soil health due to continuous intensive cropping, shrinking trend of gross cropped area and also continuous severe drought prevailed in the past 2 years, Tamilnadu State continues to occupy top place in All India ranking towards the productivity of major crops.

## Present Scenario

The present agricultural scenario are,

- Pattern of agricultural growth not uniform between region, crops and categories of farmers.
- New Technologies not permeated into dry land regions and unirrigated crops as expected.
- Number of dryland crops remain untapped to their potential
- Stagnation in growth of agricultural sector.
- Reduction of gross cropped area.
- Permanent and current fallows increased from 17.65 L.Ha. in 50's to 24.35 L.Ha. in 2001-2002.
- Depletion of water resources
- Degradation of land and soil quality
- Slow down in growth rate of major crops and yield of dry land crops.
- Yield gap exists between potential yield and actual yield of major crops.

## SCHEME ON ELITE PLANTING MATERIAL AND MODEL JATROPHA PLANTATION

The NOVOD Board have sanctioned Rs.150 lakhs (Rs.75 lakhs for Department of Agriculture and Rs.75 lakhs for TAWDEVA) for implementing the project on "Elite planting material and model Jatropha Plantation". Under this programme, the model Jatropha plantation will be established to cover 600 Ha. in the Government farms belongs to Agriculture and Horticulture Departments @ Rs.25,000/Ha. as 100% assistance towards inputs, equipments, planting materials, maintenance, etc. The Chief Engineer(Agricultural Engineering) will assist in landscaping, digging 25 bore wells and installation of Drip irrigation in these identified farms based on the need. 720 Kgs of Jatropha curcas seeds were obtained from Indira Gandhi Agricultural Uniersity Raipur and 900 Kgs. from Tamil Nadu Agricultural University, Coimbatore for raising seedlings. Besides, 25000 nos. of seedlings have been obtained from the Tamil Nadu Agricultural University, Coimbatore to take up planting in Annapannai. This programme launched on 25.2.2005 and planting commenced.

During 2005-2006 NOVOD Board will be approached to sanction of Rs.15 lakhs as maintenance subsidy for these plantations. These plantations will serve as modal plantation. The seed material to be obtained from these plantations will be utilized for further area expansion.

## PROMOTION OF SUGARBEET

The firms M/s.Mohan Breweries amd Distelleries Limited, M/s. Bannari Amman Sugars Ltd., M/s. Thiru Arooran Sugars Ltd., M/s. Shree Harichandana Estates Pvt, Ltd., have furnished to Government their expression of interest for entering into contract farming agreements with farmers in selected districts, who will cultivate Sugarbeet and sell Sugarbeet tubers to the Processing Unit of the firms concerned and Government have permitted them to take up the programme in selected districts as follows.

| Firm Name / Districts permitted by Govt. | Districts Programmed | Action Plan (acres) | Seed requirement (tonnes) | Area covered upto Feb 05 (acres) |
|---|---|---|---|---|
| M/s.Mohan Breweries & Distilleries Ltd Villupuram | Villupuram | 50 | 60 | 0 |
| | Tiruvallur | 100 | 120 | 10 |
| | Total | 150 | 180 | 10 |
| M/s.Bannari Amman Sugars Ltd., / Erode | Erode | 10 | 12 | 7 |
| | Total | 10 | 12 | 7 |
| M/s.Shree Harichandana Estates Pvt.Ltd | Madurai | The firm has no programme for 2004-05 | | |
| | Theni | | | |
| M/s.Thiru Arooran Sugars Ltd., Pudukottai, Thanjavur, Cuddalore Perambalur, Salem, Tiruvannamalai | | The firm has no programme for 2004-05 | | |
| | GRAND TOTAL | 160 | 192 | 17 |

The benefits extended by Government under Sweet Sorghum on Purchase Tax, Sales Tax on ethanol and special incentives as per industrial policy statement 2003 of Govt. of Tamil Nadu are applicable for Sugar beet also. In respect of sugar beet. it requires certain costly machineries like slicer, diffuser etc., which are not installed in the sugar factories at present. Further, the seeds are to be obtained only by import from other countries through seed companies.

Hence the firms are taking up the programme during 2004-05 only on trial basis. The programme will be extended for larger areas after studying the performance of the crop in the trials conducted during 2004-05 and after installation of required infrastructure facilities like slicer, diffuser etc., by the sugar industries.

Two firms have furnished their programme for 2004-05 to cover 160 Acres of which 17 Acres were covered. The balance 143 acres are proposed to be covered by the firms during March 2005. The promotional scheme for sweet sorghum and sugar beet will be continued during 2005-06.

## SUGARCANE

India is the second largest producer of sugar next only to Brazil in the world and contribute around 1/5th of the World's Sugar Production. Tamil Nadu ranks first in productivity of sugarcane for the past several years. Sugarcane is cultivated normally in an area of about 3 L.Ha. which constitutes about 2% of total cultivated area of the State. The average annual production of cane is 320 L.MT. The productivity of sugarcane ranges from 92.9 MT/ha. to 124.2 MT/Ha. among the districts and the average State productivity is 107 MT/ha. The scope for increasing area under sugarcane is very limited as the crop is water intensive and hence focus is mainly to sustain the production and productivity of sugarcane in the State. The following activities are carried out through Sugarcane Development Scheme in all the districts except Ramanathapuram, Thoothukudi, Kanyakumari and The Nilgiris districts.

| Sl. No. | Component | Target 2004-05 | Achmt. upto February 2005 | Programme for 05-06 |
|---|---|---|---|---|
| 1. | Release of parasite (Ha.) | 10500 | 10250 | 10500 |
| 2. | Financial (L.Rs.) | 3.75 | 3.430 | 3.75 |

## Public Sector and Cooperative Sugar Mills

Cane Cultivation and productivity for 2005-2006 of Cooperative and Public Sector Sugar Mills. The Cooperative and Public Sector Sugar Mills have proposed to register cane area of about 77600 hectares for the 2005-2006 crushing season and expected to crush about 57.50 lakh tonnes as against 32.80 lakh tonnes planned to be crushed during the current year 2004-2005 crushing season with a capacity utilization of 50%.

## Sugarcane Productivity of all Sugar mills in Tamil Nadu for 2005-2006

As the crushing operation in all the sugar mills are continuing, as on 28.2.2005, the mills have crushed 49.90 lakh MTs and at the end of the season, all the sugar mill in Tamil Nadu are expected to crush about 95.41 lakh tonnes. The average expected productivity of sugarcane is about 79 Metric Tonnes per hectare by adopting the yield improvement techniques. However, there is a scope to increase the productivity to about 100 tonnes per hectare by adopting the cane yield improvement techniques. The Government have approved an Action Plan to increase the yield in cane areas where it is less than the State average yield per hectare with the technical assistance rendered by the Sugarcane Breeding Institute and Sugarcane Research Stations.

## Action Plan to increase Sugar Recovery

An action plan has been prepared to crush 57.50 lakh MTs with a capacity utilization of 88%, the expected average yield of about 74 tonnes per hectare with a sugar recovery of 10% during 2005-2006 crushing season as against 32.80 lakh MTs of cane arranged for crushing during 2004-2005 crushing season with an average recovery of 9.97 % and 69 tonnes average yield per hectare. To achieve the above objective it has been programmed to adopt the following strategies at field level.

1. Improving the cane productivity and area under High Sugar recovery rich Varieties.
2. Training to the farmers and field staff by the Scientist of Sugarcane Research Station, to impart the new technology.
3. Arrangement for timely supply of inputs and advice the cane growers to apply ferlisers in time.
4. Adoption of drought management techniques.
5. Promotion of Drip irrigation

6. Application of Bio-fertilizers and Bio-pesticides
7. Application of Bio-compost, vermi-compost and enriched press mud.
8. To create awareness among farmers and field staff by way of furnishing the integrated pest and disease management to control the net break of new pest and diseases.
9. Multiplication of high yielding new varieties.

## Sugar Recovery and High Sugar Varieties

The high sugar varieties play a very important role in improving the sugar recovery of the mills and thereby reducing the cost of production. During 2004-2005 crushing season, the high sugar varieties to be crushed by Coop. Sugar Mills and Public Sector Sugar Mills is about 90 %, with expected recovery of 9.97%. For the next season 2005-2006 also the high sugar varieties expected to be crushed with expected recovery of 10 %. Apart from this, the quantity of cane expected to be crushed for the next season would be 57.50 lakh tonnes as compared to current season's 32.80 lakh tonnes.

## Promotion of Drip Irrigation for Cane Cultivation

Water is becoming a scarce resource and conventional irrigation methods for sugarcane involves excessive use of water and is highly labour intensive. The Government have planned to promote drip irrigation for sugarcane cultivation to conserve water and irrigate more areas. It has also been proposed to provide subsidy from Sugarcane Cess Fund to the farmers for installing drip irrigation system. For the above purpose, the Cess Fund Utilisation Committee have recommended to allot Rs.10/- lakhs from Cess Fund which is under consideration of the Government.

## Application of Bio-Fertilizer and Bio-pesticides

Bio-fertilizers like Acetobactor and Phosphobacterium are applied in the soil for fixing atmospheric nitrogen and phosphorus respectively. Application of bio-fertilizers will reduce the requirement of chemical fertilizer and thereby minimize the cost of cultivation. Bio-fertilizer units will be started in Coop. Sugar Mills and Public Sector Sugar Mills to ensure sufficient supply of Bio-fertilizers.

Steps will be taken to supply Bio-pesticides like Trichogramma and NPV in all the Public Sector Sugar Mills. Application of Bio-pesticides will make sugarcane farming economically viable and environmentally safe. Bio-control Research Laboratory at Tamilnadu Agricultural University will be strengthened to ensure sufficient supply of Bio-pesticides.

## Bio-compost

The total production of Press mud is about 2.00 lakh tonnes in all Cooperative and Public Sector Sugar Mills. Bio-compost is produced from press mud utilising the "spent wash" from the distilleries. At present there are only two distilleries- Salem and Amaravathy Cooperative Sugar Mills where the entire press mud of about 25,000 tonnes are converted into Bio-compost (i.e. 11,000 tonnes of bio compost) and supplied to the ryots.

## Vermi Compost

In other Cooperative and Public Sector Sugar Mills where distilleries are not available vermi compost is prepared by utilizing press mud and earthworms. Studies undertaken in Tamilnadu Agricultural University showed that Vermi-compost has positive impact on productivity of sugarcane and also other crops. Vermi-compost improves the soil texture and water holding capacity of the soil. Tiruttani, Cheyyar and Chengalrayan Cooperative Sugar Mills have taken up the scheme successfully. It is proposed to extend the scheme to other mills during the current year. Further, press mud is de-composed and mixed with bio-fertilizer and is distributed to the farmers as enriched press mud. All the sugar mills will be taking up this scheme and distribute enriched press mud to farmers to increase the yield and quality of cane. In the same manner cane trash is mixed with press mud and trash compost is made in the cane farm of the mills.

## Cane Price Fixation

The Government of India in their notification announced Statutory Minimum Price for 2004-05 season at Rs.745/- per MT of cane linked to 8.5% sugar recovery, subject to a premium of Rs.8.80 per tonne for every 0.1% increase in the recovery above that level.

## Collection of Cess and Sugarcane Road Development Scheme

The Sugarcane Road Development Scheme is being implemented by State Government to improve the network of link roads between the cane areas and the concerned sugar mills. The road network aims at facilitating easy movement and regular supply of sugarcane to the sugar mills. The funds for the Sugarcane Road Development Scheme are generated by levy of cess on sugarcane crushed by the sugar factories at the rate of Rs.5/- per tonne of cane purchased from outside the local area declared under the provisions of the Tamilnadu Sugar Factories Control Act, 1949. The details of cane cess collected during the last four years and expenditure on road works are given below

| Year | Sugarcane cess collected (Rs. in lakhs) | Total Expenditure (Rs. in lakhs) |
|---|---|---|
| 2001-2002 | 872.94 | 339.99 |
| 2002-2003 | 613.00 | 704.33 |
| 2003-2004 | 470.70 | 823.00 |
| 2004 - 2005 upto 28.2.2005 | 276.08 | 577.75 |

## EXTENSION ACTIVITIES

### Modernisation Of Extension System of Agriculture Department

The present system of extension is multi-layered and cumbersome. To have fast delivery system to the farmers, it is proposed to have a two tier system at the gross root level. Bearing this in mind, by providing all the infrastructure at the block level / village level, the extension wing can reach the farmers' door step. This will certainly help the farmer in receiving all the benefits from Agriculture Department which in turn certainly increase the production and productivity of the crops. The system which we are going to introduce will have a multi sectoral approach by coordinating all the service and allied sector.

### Establishment of Agricultural Facilitation Centres

It is proposed to establish 200 Agricultural Facilitation Centres in 200 non taluk headquarters blocks at a cost of Rs.100 crores through RIDF loan subject to approval by NABARD. The Agricultural Facilitation Centres will have Consultancy Room of all Departments like Agriculture, Horticulture, Animal Husbandry, Soil Conservation with Mini Laboratories , Exhibition centres and also Input godown. Providing Single Window Extension Service, bringing services closer to user and improved input supply are the major benefits of Agricultural Facilitation Centres.

### Tamilnadu Women in Agri Business And Extension (TANWABE)

It is proposed to strengthen 4000 women group in 3 years and to make them as entrepreneurs in agricultural extension activities like raising seed farm, production of seedlings, hiring of agricultural implements and also to motivate the groups to involve themselves in various programmes.

## Training Policy

Training has become an essential component in order to keep a competitive edge in the changing economic environment characterized by globalisation. Under this concept, it is aimed to identify the training needs and to conduct training on subjects related to farming technology which provide remuneration to the participants.

## ICT Enabled Agriculture

As agriculture is the most vital sector of Indian economy, the ICT enabled Agriculture envisages the many facets of the agricultural sector viz., agricultural schemes, agricultural population, rainfall, land holdings, land use pattern, irrigation, cropping intensity, crop production technologies, productivity trend, agro climatic zonal planning, agricultural commodities prices, agricultural wages and implements etc. The State Agricultural Department is responsible to provide information to the farmers on crop production technologies, assistance through various development programmes, availability of inputs and its price, market trend, etc. for sustainable agricultural development.

Under the TN-AGRISNET project it is envisaged to equip the State and District level agricultural departments with the State-of-the-art ICT infrastructures, which will facilitate agricultural related information access and advisory services to the Resource-poor-farmers in particular, and all stakeholders of the supply-chain of the Agricultural system, in general. This will ensure that the computer systems are used not only for administrative purposes but also for "farmer-centric" Information System and Access. The programme is proposed to be implemented in two major phases.

1. Thrust will be given to empower the officers and staff of Agriculture department at every level under ICT.
2. Thrust will be to provide online services to the farming community and the external agencies involved in allied services.

In the first phase, all offices up to the Block level will have to be provided with adequate computing and communication infrastructure in order to capture, process and transmit the relevant details for further processing and dissemination. The applications will have to be primarily web based for easy implementation and maintenance. Good connectivity will have to be established between the offices. While planning for wide area connectivity, the availability of TN State Wide Area Network (TNSWAN) will have to be taken into account.

The TN-AGRISNET Server Nodes are required to be established at the Commissioner Office, 28 Joint Director of Agriculture Offices at Districts, 384

Block Agriculture Development Offices, 35 Soil testing labs, 9 Pesticide Testing Labs, 6 Fertilizer Control Labs, 6 Bio-Fertilizer Production units, 12 Farmers Training Centres, STAMIN, 37 State Seed Farms, 5 State Oilseed farms and 1 Pulses farm.

This ICT activities in agriculture is proposed to be taken up with RIDF to the tune of Rs.19.06 crores loan subject to approval of the project by NABARD. Besides, an amount of Rs.5 lakhs is provided under Part II scheme to augment the ICT facilities to farmers linking (Rural Access Service through Internet) RASI Project during 2005-06 .

## Web Based Data Collection for the Department

To ensure effective monitoring of the crop coverage, scheme progress and also administrative aspects, Monday Message Monitoring System on weekly basis, Personnel Information System and File Monitoring System have been established at Commissionerate of Agriculture. This Web based Data collection helps to speed up the activities of the Department besides creating a Data Bank for future planning.

## State Agricultural Extension Management Institute (STAMIN)

To train the extension personnel of the Department of Agriculture and to equip them with upto date technologies, approaches and also to enhance their managerial ability, a State Level Training Institute (STAMIN) is functioning at Kudumianmalai. The STAMIN is also engaged in conducting special training courses to NGOs, Agricultural Students, personnel from Sister Departments. The details of training courses conducted during 2004-05 are given below.

| Sl. No. | Name of the Training | No. of Officers trained | |
|---|---|---|---|
| | | Target | Achievement upto Feb. 05 |
| 1. | Office Administration Training | 200 | 140 |
| 2. | Office Management Training | 220 | 160 |
| 3. | Computer Training | 100 | 80 |
| | TOTAL | 520 | 380 |

## Farmers Training Centres

Dissemination of latest technologies to the farmers and ensure adoption is the prime responsibility of the Department. Training of farmers is a continuous process due to the change in agricultural scenario like cropping pattern, market preference, demand for agricultural produce, value addition, crop diversification as these are deciding factors to get profit in Agriculture. With the objective 12 FTCs are functioning in the state. These are located at Kancheepuram, Tindivanam, Vellore (originally functioned at Navlock), Salem, Dharmapuri, Erode (originally functioned at Bhavani) Lalgudi, Kudumianmalai, Sakkottai, Paramakudi, Palayamkottai and Nagercoil (originally functioned at Tirupathisaram). Besides imparting training at village levels, training cum demonstration, method demonstration and farmers tours are also conducted in collaboration with TamilNadu Agricultural University. Farmers day is also celebrated wherein lot of Farmers participate and share their experience. Extension activities like farm news, through Radio, Newspaper and News letter are also carried out, to ensure speedy transfer of technologies. Target and achievement under Farmers Training Centre:

| Sl. No | Particulars | Unit | 2004-05 | | 2005-06 |
|---|---|---|---|---|---|
| | | | Target | Achmt. upto Feb 2005 | Programme |
| 1. | Village based training for men & women | Nos. | 264 | 256 | 264 |
| 2. | Village based training for FDG conveners | Nos. | 120 | 117 | 120 |
| 3. | Field demonstration | Nos. | 1320 | 1266 | 1320 |
| 4. | Method demonstration | Nos. | 528 | 510 | 528 |
| 5. | Financial Achievement (L.Rs.) | | 6.57 | 4.995 | 6.57 |

Considering the importance in training the farmers on latest technologies and to create opportunities for the farmers of all districts to get training benefits, the Hon'ble Chief Minister has ordered to establish new Farmers Training Centres in 11 districts. The 10 FTCs to be established at Thiruvannamalai, Namakkal, Karur, Perambalur, Dindigul, Theni, Sivaganga, Virudhunagar, Thoothukudi and Krishnagiri districts will functrion under Agriculture Department and one FTC at Nilgiris under the control of Horticulure Department during 2005-06 where the Farmers Training Centres belongs to the Department and Krishi Vigyan Kendras belongs to Tamilnadu Agricultural University are not functioning now. The 11 new Farmers Training Centres will be established during 2005-06 at a total cost of Rs.100 lakhs.

## STUDY ON EXPORT PROMOTION

The WTO Agreement poses threats and opportunities. It is imperative that a detailed study and consultancy services is essential on Crop competitiveness both domestically and internationally and also the constraints in both forward and backward linkages to device modalities/ strategies to provide relevant recommendations to farmers and for proper implementation.

Under Ministry of Commerce, Government of India, a Market Access initiative (MAI) scheme is in operation wherein assistance are extended to supplement the State Government in carrying out export potential surveys of the States for identified product groups to evolve market related strategy for promoting exports of the identified product groups from the State. 50% of the cost of Survey will be provided as GOI assistance. Hence this study will be taken up on sharing basis at a cost of Rs.20.00 lakhs. (State Rs.10.00 lakhs + GOI (MAI scheme) Rs.10.00 lakhs) during 2005-06.

## Crop Yield Competition

Crop Yield Competitions are conducted at the State level as well as at the District level to motivate the farmers to achieve highest productivity by adopting the advanced scientific techniques in crops like paddy, groundnut, cholam, cumbu, greengram and blackgram and cash prizes are awarded. The prize amounts given are as follows:

| Crop | State Level (In Rs.) | | District Level (In Rs.) | |
|---|---|---|---|---|
| | 1st place | 2nd place | 1st place | 2nd place |
| Paddy and groundnut | 25000 | 15000 | 8000 | 4000 |
| Other crops | 8000 | 4000 | 4000 | 2000 |

The enrolment fee for paddy and groundnut crops for State level competition is Rs.100 and for other crops is Rs. 50 /-.

Similarly, the enrolment fee for paddy and groundnut crops for District level competitions is Rs.50/- and for other crops it is Rs. 25/-. This programme is being implemented in 2004-2005 at a cost of Rs15.26 lakhs. This programme will be continued during 2005-2006.

## CENTRALLY SPONSORED SCHEMES

### Schemes Under Macro Management

The Government of India have launched the Macro Management Mode schemes in October 2000. This Macro Management Mode Scheme is a major step towards decentralization/ restoring primary role of the State in agricultural development planning.

Under Macro Management Mode, 27 identified Centrally sponsored Schemes were amalgamated and the States have the flexibility to develop programmes based on regional priorities through work plan comprising integrated approach on activities of agriculture, horticulture, Land Development, Soil Conservation, Marketing, Cooperation and State Land Use Board.

The expenditure under this scheme is shared by Government of India and State on 90:10 basis. Out of 90% of the GOI share, 80% is extended as grant-in-aid and balance 20% is treated as loan.

Under Macro Management Mode Scheme the following schemes are implemented by Agriculture Department.

### Sugarcane Development Scheme under Work Plan

Government of India has approved inclusion of SUBACS (Sustainable Development of Sugarcane Based Cropping System) under Macro Management Mode Scheme.

The scheme is implemented in all the districts except Ramanathapuram, Thoothukudi, The Nilgiris and Kanyakumari. Training of farmers on improved cultivation technologies and release of parasite for control of sugarcane borer is contemplated.

Drip irrigation part of the scheme is implemented by the Agricultural Engineering Department. The progress under this scheme is as follows:

| Sl. No. | Component | Unit | Physical | | Financial (L.Rs.) | |
|---|---|---|---|---|---|---|
| | | | Target 2004-05 | Achmt. upto Feb 2005 | Target 04-05 | Achmt. upto Feb 2005 |
| 1. | Farmers Training assistance @ Rs.5000/ training. | Nos. | 100 | 97 | 5.0 | 4.850 |
| 2. | Release of Parasite @ 25% limited to Rs.10/Ha. | Ha. | 5000 | 4634 | 0.50 | 0.463 |
| | TOTAL | | | | 5.50 | 5.313 |

This scheme will be continued during 2005-06.

## Balanced and Integrated Use of Fertilisers

The programme on Balanced and Integrated Use of Fertiliser is implemented under Macro Management Mode Scheme. The activities taken up under this programme during 2004-05 are:

### *Conducting Orientation Courses to Technical Personnel*

Training is extended to 360 technical Officers on Micro Nutrient Analysis, uses of biofertilisers, organic farming, reclamation of problem soils and their management, etc. at Central Control Laboratory, Kudumianmalai in 12 batches at 30 Officers per training at a cost of Rs.1.20 lakhs during 2004-05.

### *Distribution of Soil Health Cards*

The Soil Health Card is a semi permanent record containing details of recommendations on fertilisers, organic manure to be applied for the prescribed crop, besides recommendation on water quality, information on soil problem and relevant reclamation measures. During 2004-05, one lakh nos.of Soil Health Cards are being distributed at a cost of Rs.4.50 lakhs.

### *Distribution of Leaf Colour Chart*

36,200 nos. of leaf colour charts is being distributed to the farmers with the leaflet containing the recommendation. The farmers can compare the chart with the leaf colour of the standing crop in their field to know the 'Nitrogen' need of crop and can apply the recommended quantity. This leaf colour chart will helps the farmers to apply optimum dose of 'N' by assessing the actual need of the crop and to avoid excess application of 'N' to the crop to reduce unproductive expenditure on fertilisers.

### *Printing of Soil Atlas*

Soil and land based agricultural planning has been gaining momentum for increasing the agricultural production. Soil is the basic non renewable natural resource. Hence, printing of soil atlas for the left over 5 district of Villupuram, Cuddalore, Thiruvannamalai, Virudhunagar and Thoothukudi has been taken up at a cost of Rs. 4.8 lakhs during 2004-05. This Soil Atlas will contain various information like geology, geomorphology, land use pattern, land capability, land / crop suitability, soil depth, colour, slope, drainage, etc. The Soil Atlas provides holistic information at Taluk/ District level for Planner, Researcher, Development Officer, Field Functionaries and to a certain exigent directly to the farmers also.

## Centrally Sponsored Scheme on Intensive Cotton Development programme under Mini Mission II of Technology Mission on Cotton

Technology Mission on Cotton was launched in February 2000 and become operartive from 2000-01 under Mini Mission II in Tamil Nadu with a aim of increase the production of cotton by providing key inputs, Demonstrations, Trainings and effective Plant Protection activities through Integrated Pest Management approach. The scheme is shared between central and state on 75:25 basis and implemented in all the districts of the State except Nilgiris, Kanyakumari, Thiruvallur and Kancheepuram.

During 2004-05, an amount of Rs.297.50 lakhs have been sanctioned for the implementation of the programme and further an additional amount of Rs.152.55 lakhs have been approved by Government of India totalling the scheme outlay to Rs.450.05 lakhs.

## SEASONAL CONDITION AND CROP PROSPECTS

The health, growth and normal yields of crops depend upon certain optimum conditions of rainfall, temperature, wind, cloudiness, etc., which normally occurs in a season. Magnitude of vulnerability of agriculture within a season again depend heavily upon weather variability within a season.

The Agricultural situation in Tamil Nadu largely depends on the quantum of rains received during seasonal rainfall of South West and North-East monsoon. Comfortable storage in Mettur reservoir coupled with uniform distribution of monsoon rains will bring good crop in delta region. But the State had faced consecutive drought situation for the past 2 years because of low and uneven precipitation.

During 2004-05, the state had experienced a dual situation, whilst it faced severe drought upto June-July 2004, it had torrential rains in the month of October leading to flood damage and submergence in some of the districts. The water from Mettur reservoir was released on 12.8.2004 after a delay of 2 months for Samba cultivation.

The area coverage under major crops showed a dip in coverage in the early part of the year, which however revived in the month of September 2004 onwards.

***The details of rainfall received during 2004 is as follows. (Unit: in mm)***

| Sl. No | Season | Normal Rain-fall | Actual Rainfall | | Deviation (%) w.r.to | |
|---|---|---|---|---|---|---|
| | | | 2003 | 2004 | Normal | Last Year |
| 1 | Winter (January-February) | 37.5 | 8.7 | 12.1 | (-) 67.7 | (+) 39.1 |
| 2 | Summer (March-May) | 127.5 | 124.9 | 283.4 | (+) 122.3 | (+) 126.9 |
| 3 | Southwest Monsoon (June-September) | 324.3 | 329.7 | 459.8 | (+) 41.8 | (+) 39.5 |
| 4 | Northeast monsoon (October-December) | 464.7 | 404.5 | 470.9 | (+) 1.3 | (+) 39.5 |
| | Total | 954.0 | 867.8 | 1226.2 | (+) 28.5 | (+) 41.3 |

On 26th December, 2004 Tsunami waves which is unusual in the India Ocean attacked Tamil Nadu Coastal line and made fertile lands submerged with sea water. The crops raised in 4,796.10 Ha. were affected and the extent of area to be reclaimed is assessed as 3,664.24 Ha. and the total extent damaged in agriculture is 8,460.34 Ha.

For relief to the crop damage, the Government have already ordered for the financial compensation of Rs.1.80 crores and a sum of Rs.16.96 crores towards the relief work to be taken up for reclamation of Tsunami affected agricultural lands.

The Mobile Soil Testing Laboratories were deputed to Nagapattinam, Cuddalore and Kanyakumari districts to make preliminary assessment of soil damage. The damage caused by Tsunami are so complex and site specific and hence 100% soil survey of the damaged land in Nagapattinam was ordered and carried out.

The Experts recommended soil amelioration by scrapping and removal of silty clay deposits, leveling, leaching with good water 3 or 4 times with short intervals to restore soil health, application of gypsum at 2 tonnes per hectare, liberal addition of organic manure or green manure. Mulching, raising salt tolerant crop/ varieties with 20% extra seed rate to ensure better germination.

## Rabi Campaigns

In the era of globalization, Indian agriculture will face increasing competition.

During initial period of Green Revolution agricultural extension played a crucial role in carrying the messages to the various parts of the State. However, with the increasing complexities in market and society, the extension methodology has to be sensitized to the new dimensions in agriculture. The task of extension is becoming more challenging in the context of globalization, which demands a system of market-linked extension. New approaches like precision farming knowledge management with modern information and communication technology are tools, which have to play crucial roles in strengthening sustainable agriculture of Tamilnadu and prepare it for facing competitive market.

Continuous drought for the past 2 years and also during early part of Kharif 2004 for the third year badly affected the agricultural situation in Tamilnadu. The gross cropped area declined to 53.16 L.Ha. from the average normal coverage of 63 L.Ha. leaving about 9.8 L.Ha. under fallow. The ground water level gone down to critical level and almost all the tanks and reservoirs went dry and posed threat to Agriculturists.

The receipt of adequate wide spread rains in late part of Kharif 2004 have created a conducive favourable situation for crop cultivation. Taking advantage of this situation, the department had taken sincere efforts to maximize the area coverage by creating enthusiasm among the farmers to take up cultivation through intensive campaign in all villages, at different stages.

- Immediately on release of Cauvery water in August, 2004, District Facilitation Centres were opened in the Delta districts and kept opened from 8.00 AM to 8.00 PM from 15.9.2004 and again 9.11.2004 to render all required assistance. Inputs were made available on war footing manner to bring maximum area under Samba. Intensive Preseason Campaigns were conducted in all the villages of delta districts.
- Phase I Intensive Rabi Campaign conducted in all the districts covering all the 17244 villages and 21029 campaigns conducted from 16th September 2004 to 15th October 2004. 4,55,336 farmers participated and benefited by these campaigns.
- Phase II campaign conducted at FIRKA LEVEL from 16th November 2004 to 31st December 2004 involving all line Departments such as Horticulture, Agricultural Engineering, Animal Husbandry, etc. besides Credit Institutes, leading input supply agencies involving people's representatives with mass media support. Inputs were made available at farmers doorstep to the extent possible. In 1086 firkas, 1750 Nos. of Phase II campaigns conducted benefiting to 92209 farmers.
- Phase III campaign conducted from 16th January,2005 to 28th February, 2005.

The Preseason campaigns conducted at different phases based on the seasonality has really had very good impact among the farmers and welcomed by all. The Department also ensured supply of quality inputs to the farmers in time. These campaigns pave the way to the farmers to adopt new technology to get good yield.

The area coverage and estimated production upto 21.3.2005 under important crops are indicated below:

| Sl. No. | Crops | Area (L.Ha.) | | Production (L.MT.) | |
|---|---|---|---|---|---|
| | | Target | Actual | Target | Actual |
| 1 | Rice | 20.00 | 19.28 | 75.00 | 69.02 |
| 2 | Millets | 11.00 | 12.12 | 14.85 | 15.20 |
| 3 | Pulses | 7.50 | 8.46 | 3.38 | 3.66 |
| | Total food-grains | 38.50 | 39.86 | 93.23 | 87.88 |
| 4 | Cotton (L.bales) | 2.50 | 1.63 | 5.62 | 2.81 |
| 5 | Sugarcane (Gur) | 3.00 | 2.37 | 35.00 | 25.29 |
| 6 | Oilseeds | 10.00 | 8.64 | 17.00 | 13.91 |

Anticipating a favourable situation, the target proposed for 2005-06 is given in the Table below.

| Sl. No. | Crops | Area (L.Ha.) | Production (L.MT) |
|---|---|---|---|
| 1 | Rice | 20.00 | 78.00 |
| 2 | Millets | 11.00 | 15.51 |
| 3 | Pulses | 8.00 | 3.76 |
| | Total Food-grains | 39.00 | 97.27 |
| 4 | Cotton (L.bales) | 2.50 | 5.85 |
| 5 | Sugarcane (Gur) | 3.00 | 36.40 |
| 6 | Oilseeds | 10.50 | 18.59 |

## Strategies

The strategies of the Government to achieve the goal in Agricultural sector are;

- Adoption of alternative cropping strategies derived for agro-climatic zone based cropping pattern evolved by TNAU to raise remunerative crops and to ensure maximum utilization of available land and water aiming to increase cropping intensity.
- As a part of Alternative Cropping Strategy, cultivation of Jatropha, Sugar beet and Sweet sorghum are aiming at, as announced by the Hon'ble Chief Minister on Contract farming basis through approved Industrial entrepreneurs for the production of Ethanol and Bio diesel. These crops are highly remunerative and can be grown in moderately fertile lands with comparatively less water.
- Taking up cultivation in vast tract of waste lands through Massive Comprehensive Waste Land Development programme as announced by the Hon'ble Chief Minister.
- Much emphasis on dry land development integrating Watershed Development and Waste land Development programme
- Adoption of dry land development technologies and crop production technologies for dry land crops to step up the productivity.
- Identification and promotion of relevant technologies to bridge the gap between the potential yield and actual yield of major crops.

- Speedy transfer of technologies to the farming community through ICT (Information and Communication Technology).
- Conduct of Seminars, Workshops and Intensive Pre-season campaigns involving line Departments to enthuse and assist farmers to harvest good crops and to get good profit.
- Efforts to provide technologies and advice on Agriculture, Horticulture soil and moisture conservation, Animal Husbandry etc., under one roof.
- Introduction of Contract farming system for Maize, Oilseeds, Pulses and Cotton in potential districts with assured buy back arrangements at pre-announced price or prevailing market price if it is high. The main objective of the programme is to ensure expansion of area under these crops and to motive farmers especially SF / MF farmers to get assured remunerative price for the produce.
- Ensure availability of quality seeds on enhanced SRR through Public Private Partnership.
- Massive adoption of integrated nutrient management and integrated pest management technology with emphasis on ecofriendly Agriculture Development.
- Much focus on restoration of soil through organic farming approach and to promote vermi composting, compost making through plueorotus, green manuring etc.
- Efforts to provide site specific macro and micro nutrient recommendations on the basis of soil test villagewise fertility Index.
- Promotion of micro irrigation to maximize water use efficiency.
- Empowerment of women by revitalizing TANWA groups.

## STATE SCHEMES

### Input Programmes

#### *Seeds*

Seed is a vital input that plays a critical role in increasing agricultural productivity. The productivity of crop largely depend on the genetic and physical purity of seeds. The Department of Agriculture has the primary responsibility of producing and distributing good quality certified seeds as per approved seed replacement rate. The Department of Agriculture has taken special initiative to address the following key areas to revitalize the quality seed production and supply.

- To achieve enhanced SRR through Public Private partnership

- To achieve National standards of S.M.R
- Proper utilization of valuable Breeder / Foundation seeds to enhance 'C' seed production.
- Promote seed village concept
- Focus to increase availability of quality seeds of dry land crops for area expansion and productivity increase.
- Strict enforcement of quality parameters for seeds
- Steps to enhance the efficiency of delivery mechanism
- Identification and introduction of High yield potential varieties released by other states / ICAR which are notified and suitable for Tamil Nadu conditions.
- Ensure advance planning on demand and supply by creating linkage with Department, Quasi Govt. Private seed producers.
- Strengthening of infrastructure for quality seed production and storage.
- Ensure all information on the portal such as availability, sale price, seed production / storage technologies, Data base on seed producer, seed dealers.

The following SRR are proposed to be adopted during 2005-2006, for Departmental Seed Multiplication Programme.

| Crop | Seed Replacement Ratio |
|---|---|
| Paddy | 17% |
| Millet | 7% |
| Pulses | 12.5% |
| Cotton | Rainfed 10% Irrigated 15% |
| Oilseedsa) Groundnut | 5% |
| b) Gingelly | 15% |
| c) Sunflower | 50% |
| d) Castor | 30% |
| e) Soya | 20% |

The main objective of the Seed Multiplication schemes undertaken by the Department are to produce and distribute Quality seed of Paddy, Millet, Pulses, Oilseeds & Cotton in time at reasonable price. The trading activity by the Department also helps to curb unnecessary price hike by the private seed traders.

The seed production activities are carried out in 3 stages.

i. Breeder Seed production by TNAU for State varieties and ICAR Centre for national varieties.

ii. The foundation seed production in 37 State Seed Farms, 5 State Oil Seed Farms and one pulses farm at Vamban.

iii. Certified seed production in farmers' holdings.The Department of Seed Certification is involved to ensure quality of seeds.

The Foundation and certified Seeds produced in Government Seed Farms and farmers' holdings respectively are processed in 16 Major 2 Medium and 56 Mini Seed Processing Units in the State with annual capacity of 28500 MT.

The entire seeds produced are distributed to the farmers through 880 Agricultural Extension Centres located in the State.

The Quantity Seeds of Paddy, Millet, Pulses, Cotton and Oilseeds procured and distributed upto February 2005 and the target proposed for 2005-06 is as follows:-

(In MT)

| Crop | 2004-2005 | | Programme for 2005-2006 |
|---|---|---|---|
| | Target | Achievement | |
| **A. Procurement** | | | |
| 1. Paddy | 17000 | 11341 | 17000 |
| 2. Millets | 400 | 340 | 400 |
| 3. Pulses | 1900 | 1064 | 1900 |
| 4. Oilseeds | 8442.3 | 2562 | 8708 |
| 5. Cotton | 367 | 125 | 367 |
| **B. Distribution** | | | |
| 1. Paddy | 17000 | 10465 | 17000 |
| 2. Millets | 400 | 340 | 400 |
| 3. Pulses | 1900 | 1181 | 1900 |
| 4. Oilseeds | 8442.3 | 2933 | 8708 |
| 5. Cotton | 367 | 137 | 367 |

The Government have made substantial allocation for implementing seed multiplication schemes and the details are as follows:-

(Rs. in lakhs)

| | | | (Rs. in lakhs) |
|---|---|---|---|
| Crop | BE 2004-05 | RE 2004-05 | Achmt. upto February 2005 |
| Paddy & Millet | 1953.44 | 2063.460 | 1601.691 |
| Pulses | 420.06 | 401.40 | 250.586 |
| Oilseeds | 602.34 | 601.140 | 515.250 |
| Cotton | 119.71 | 93.650 | 61.480 |

During 2004-2005, as announced in the Assembly, 100 Agricultural Extension Centres were repaired at a cost of Rs.100 lakhs for proper safe stocking of seeds to ensure distribution of quality seeds to the farmers.

For drying and also for protecting seeds, 213 Nos. of tarpaulins were purchased at a cost of Rs.10 lakhs.

## Fertilizer

The Department of Agriculture shoulders the responsibility to closely monitor the production and supply of fertilisers to ensure timely availability to the farmers. After continuous drought situations for the past more than 2 years, during 2004-2005, extensive cultivation was resorted to by the farmers due to favourable seasonal rains from September 2004. This has created greater demand for fertilizers especially urea. This sudden spurt in demand was effectively tackled by securing additional quantity of 2.38 L.MT of urea during Rabi till November 2004 despite extreme limitations and timely availability ensured. A public notice was released in the dailies sternly warning the traders against indulging in over pricing or anyother malpractice and also inviting complaints from the farmers on non availability, sale at higher price. To enforce this warning 608 fertilisers shops were inspected and 79 stop sale orders issued 42 nos.of warning issued and 59 nos.of license suspended temporarily. To monitor the fertilizer supply, Facilitation centres were opened in all the districts from 8.00 A.M to 8.00 P.M. and the Department ensured timely availability of fertilizer at correct price.

The progress on supply of chemical fertilisers during 2004-05 and programme for 2005-06 are as follows:-

(Unit in LMTs)

| Nutrient | Target 2004-05 | Achievement upto February 2005 | Programme for 2005-06 |
|---|---|---|---|
| Nitrogen | 5.40 | 5.18 | 5.40 |
| Phosphorus | 2.31 | 2.26 | 2.31 |
| Potash | 2.24 | 2.38 | 2.24 |
| Total | 9.95 | 9.82 | 9.95 |

## Micronutrients

Micronutrients are very important for maintaining the Soil Health and also for increasing productivity of crops. Micro Nutrient deficiency is wide spread as evidenced in the increase in the percentage of Soil exhibiting deficiency of micro nutrient over years. Poor adoption of Micro nutrient Management is another reason for stagnation of Agricultural Productivity. The adoption of Intensive cultivation practices causes depletion of Micronutrient levels in most of the soil. The Department of Agriculture is constantly advocating balanced application of Micronutrients to the soil suitable for different crops. 18 Atomic Absorption Spectro Photo meter (sophisticated equipment to test micro nutrient content in the soil) have already been provided in the Soil Testing Laboratories. To meet the soil testing requirement of Sivagangai and Virudhunagar, during 2004-2005, 2 Nos. of Atomic Absorption Spectro Photometer were purchased for the above 2 districts also and by this, all the districts have been extended with facilities for testing the soil for Micro nutrient content. To make available quality Micro Nutrient mixture through the Departmental outlets, one Micro Nutrient Production Unit is functioning at Kudumianmalai to produce and supply 1250 MT of Micronutrient mixtures annually for Paddy, Millets, Pulses and Oilseed crops.

During 2004-05, upto February 2005, 1192 M.T of Micronutrient mixture have been produced and 1094 MTs were distributed. This programme will be continued during 2005-06 also.

## Fertilizer Control Laboratories

The Govt. is very keen to make available quality fertilizers to the farmers as per the standards prescribed under the Fertilizer Control Order, 1985.

Strict enforcement work are being done by drawing samples from wholesale and retail sale outlets as well as from the mixing units. There are 14 notified Fertilizers Control Laboratories functioning in the State, with the analyzing capacity of 16,800 samples per annum. During 2004-2005, upto February 2005, 15,603 Nos. samples were analyzed. During 2005-2006, 16800 Nos. of fertilizer samples will be analysed by the laboratories to ensure quality.

## Organic Farming

Due to continuous cropping, indiscriminate use of chemical fertilizers and inadequate application of organic manure to the soil, the general soil health of the State is getting deteriorated. The organic manure content in the soil has gone down from 1.20% in 1971 to 0.68% in 2002 in Tamilnadu, because of less use of organic inputs.

The decline in organic matter content has made undesirable changes in soil biodiversity and disruption in harmony of crop plants, which affects soil fertility and productivity.

Considering its importance, the application of Bio-fertilizer, Green Manure, Green Leaf Manure, Vermi Composting, Composting of farm wastes through Pleurotus is popularized as a part of Integrated Plant Nutrient Management Technology. To ensure the adoption of the organic farming technology the following programmes are implemented.

### *Composting of Farm Waste through Pleurotus*

During 2004-05, 2000 Nos. of kits each containing 1 Kg of Pleurotus, 5 Kgs of Urea and one Pamphlet have been distributed to the farmers at free of cost with an expenditure of Rs.2.80 lakhs to make compost by utilizing farm waste through Pleurotus for field applications. This programme for the distribution of 2000 No. of kits at a cost of Rs.2.80 lakhs will be continued during the year 2005-06 also.

### *Scheme to Promote vermi Composting*

Under Vermi Compost Scheme, 100 Nos. of demonstration, cum trainings benefiting totally 5000 farmers at a total cost of Rs.3.85 lakhs has been conducted during 2004-05. The Scheme will be continued during 2005-06 also. The Department is also taking efforts to produce vermi compost in the Govt. farms for distribution to the farmers to the extend possible through AECs.

### *Biofertilizer Production and Distribution*

To increase the availability of Bio-fertilizers, the Production of Bio-fertilizer inoculum packets containing Rhizobium, Azospirillum and Phosphobacterium are taken up in 6 Biofertilizer Production Units functioning at Cuddalore,

Ramanathapuram, Salem, Kudumianmalai, Sakkottai and Trichirappalli with the Annual Production capacity of 1400 Mts.(70 L.Nos. of pockets of 200 gms each).

The quality of biofertilizers produced in these units is monitored through Biofertilizer quality control Laboratory functioning at Trichirappalli.

The Biofertilizer Production Programme for 2005-06 is as follows:

| Sl.No | Name | Quantity (MT) |
|---|---|---|
| 1 | Rhizobium | 280 |
| 2 | Azospirillum | 800 |
| 3 | Phosphobacterium | 320 |
| | Total | 1400 |

During 2004-2005 upto February 2005, a quantity of 1109 MTs biofertiliser were produced and 940 M.T. were distributed to the farmers through Agricultural Extension Centres. This programme will be continued to produce and supply 1400 M.T during 2005-06 also.

## Blue Green Algae and Azola

The Blue Green Algae and Azolla are reported to act as a supplementary sources of Nitrogen and have the capacity to fix atmospheric Nitrogen and made available to the crop. To make the inoculum available to the farmers, 500 M.T of Blue Green Algae and 500 M.T of Azolla are produced in the Govt. farms and supplied to the farmers.

### *Scheme to Promote Green Manuring*

Raising of Green manure crops and ploughing insitu will enhance organic content of the soil. This practice has suffered by the intensive cultivation of other field crops.

As there is considerable decrease of animal husbandry in the farms, the availability of Farm yard Manure is felt inadequate and hence use of green manuring and green leaf manuring are the alternative source that could add organic matter to the soil.

Hence to motivate and to encourage the farmers to raise the Green manure crop, a state scheme is under implementation at a cost of Rs.25.00 lakhs to produce 125 M.T of Green Manure seeds and to distribute at 25% subsidy.

During 2004-05, 137 M.T of green manure seeds have been produced upto February 2005 and 117 M.T Green Manure Seeds have been distributed at 25% subsidized cost.

This programme will be continued during 2005-06 also.

### *Popularising INM (Integrated Nutrient Management) technologies among farmers through Farmers Training Centres*

Resource exhaustive crop varieties popularized during the Green revolution phase have brought in question of soil nutrient sustainability. While major nutrients are applied to soil through inorganic fertilizers, continuous mono-cropping takes up cumulatively higher quantities of other minor nutrients that are not replenished.

In the absence of organic manure application, the soil profile is altered. Application of large doses of chemical fertilizers in an imbalanced way not only polluted environment and soil but also not given economic returns in terms of yield. Hence to keep soil health and productivity adopting Integrated Nutrient Management technology has become highly essential. This technology is being popularized through training programme to farmers in the 12 farmers training centres.

This training programme was introduced to train 2700 farmers during 2003-2004 at a cost of Rs.8.64 lakhs. This training programme have been enlarged during 2004-2005 and totally 13500 farmers were trained at the rate of 500 farmers per district at a cost of Rs.43.20 lakhs. Each training was conducted for 2 days and 50 farmers were trained per batch. Totally 270 trainings were conducted.

The achievement is as follows:

| Sl. No. | Details | Target | Achmt. upto February 2005 |
|---|---|---|---|
| 1. | INM - No.of training conducted (Nos.) | 270 | 270 |
| 2 | No.of farmers trained | 13500 | 13500 |
| 3. | Financial (Lakhs) | 43.20 | 43.20 |

## SOIL MANAGEMENT AND SOIL TESTING

The objective of the scheme is conducting soil survey and preparation of soil resources inventory including the nature of soil occurring in the area, their morphological, physical and chemical characteristics through field studies and laboratory analyses, classifications according to internationally recognized system of soil classification and mapping their extent on standard topographic base maps and finally interpreting for variety of uses.

The above survey is being taken up through the four Soil Survey units at Coimbatore, Thanjavur, Vellore and Tirunelveli.

During 2004-05 an area of 0.99 L.Ha. has been surveyed as against the target of 1.10 lakh ha. till February 2005. An area of 1.10 L.Ha. will be covered by "detailed soil survey" by these units during 2005-2006.

## Reclamation of Saline and Alkaline Lands

The Saline and Alkaline Land Reclamation Scheme is functioning to reclaim problem soils in Nagapattinam, Thiruvarur, Kancheepuram, Thiruvallur, Vellore, Thiruvannamalai, Salem and Namakkal districts in an area of 4000 ac. During the year 2004-2005, an area of 2886 acres were reclaimed up to February, 2005. It is programmed to reclaim 4000 acres during 2005-2006 under this scheme.

## Soil Testing Laboratories

As a part of Integrated Nutrient Management technology, the farmers have to be motivated to apply need based Macro / Micro nutrients to the crops, to ensure economy in input use for higher productivity and profitable return.

To advise the farmers on site specific fertilizer recommendations on the basis of native soil fertility, analysis of soil samples is being carried out through 19 Soil Testing Laboratories and 16 Mobile Soil Testing Laboratories functioning in the State.

A total number of 4.13 Lakh of soil samples have been analyzed through Soil Testing Laboratory against the target of 5.28 L.Nos. and a total number of 1.71 Lakh Nos.of soil samples through Mobile Soil Testing Laboratory were analysed till February 2005 against the targetof 2.82 L.Nos.

The above laboratories also suggest suitable reclamation measures for the problem soils, preparing village level fertility index and assessing the quality of irrigation water.

The Mobile Soil Testing Laboratories are the effective propaganda aids to disseminate the soil, bio-fertilizers and micronutrient management technologies among the farming community through Village Adoption Programme, Soil Testing campaigns, Exhibitions, etc.

Moreover, there is an apex organization viz. "Central Control Laboratory" located at Kudumianmalai, which inculcates technical competence through training to the laboratory personnel and also monitors accuracy of analysis of all the laboratories in the Department of Agriculture.

## PLANT PROTECTION

### Pest Surveillance

The Integrated Pest Management Technologies are recommended and adopted extensively by the farmers. The success of the Integrated Pest Management largely depends on constant monitoring of pest and disease situation in the field, defender population, etc.

To augment its efforts the pest surveillance are done in 7356 locations every week and forewarning messages are given to the farmers on the possible out break. Because of the continuous surveillance, the serious pests like paddy stemborer, leaf roller, Brown Plant Hopper, groundnut red hairy catter pillar, prodenia, cotton boll worm, etc. which caused severe damages a decade ago were kept under control.

### Crop and Plant Protection Scheme

The Plant Protection measures were programmed to be taken up during 2004-05 to cover an area of 37.5 lakh ha. in food crops, 18.2 lakh ha in non food crops and seed treatment measures in 22.0 lakh ha. and upto February, 2005 an area of 31.760 Lakh ha. were covered in food crops, 15.132 Lakh ha. under non food crop and 22.693 L.Ha. under seed treatment. This will be continued during 2005-06 also.

Under this scheme, Government have sanctioned funds for the procurement of bio-control agents and bio-pesticides for distribution to the farmers at subsidized cost through Centrally sponsored Schemes such as Cereals Development Programme, SUBACS, ICDP Cotton, ISOPOM-Pulses, Oil Seeds and Maize.

### Bio-control Agents Production Centre & State IPM Centre

As a part of IPM activities, production of Bio control Agents are resorted by the Department and distributed to the farmers.

There are 12 State Bio-Control Agents Production Centres and 51 Parasite Breeding Centres functioning in this State to take the production of the Bio Control Agents as indicated below:

| Sl. No. | Bio control agent / Crop / Pest | No .of Centres | Area Coverage (In Ha.) |
|---|---|---|---|
| 1 | Trichogramma Chilonis (Sugarcane Internode borer) | 21 | 10500 |
| 2 | Bethylids, Branchonids and Eulophids (Coconut black headed caterpillar) | 21 | 10250 |
| 3 | NPV (Cotton Boll Worm) | 3 | 1500 |
| 4 | NPV (Groundnut red hairy Caterpillar and Prodenia) | 4 | 480 |
| 5 | Green Muscardine fungus tubes (Coconut Rhinocerous beetle.) | 2 | 10000 |
| 6 | Bio-control agents (Cotton, Paddy, Pulses, Groundnut) | 12 | 60000 |

This scheme will be continued during 2005-06 also.

## Integrated Pest Management (IPM)

The Integrated Pest Management Technology to combat pest and diseases was introduced in Tamilnadu through FAO Programme in late 80's and special thrust were laid to this concept through Farmers Field School approach.

This season long training cum demonstration component involving 30 farmers in each training, created greater awareness among the farmers and being practiced in almost all the villages.

Because of the large scale adoption of IPM technologies, the technical grade pesticide consumption which was annually around 10,000 MTs. a decade ago has come down to around 1300 MTs.

During 2004-05 the following programmes are implemented under IPM through Centrally Sponsored Schemes.

| Sl. No. | Name of the component | Physical (Nos.) | |
|---|---|---|---|
| | | Progr-amme | Achievement upto Feb 05 |
| 1 | IPM- Demonstration cum FFS under Cereals Development Programme | 330 | 330 |
| 2 | IPM Demo cum FFS under Cotton Mini-mission II | 350 | 340 |
| 3 | IPM Demo cum FFS under ISOPOM maize | 16 | 16 |
| 4 | IPM Demo (FFS) under ISOPOM - Oilseeds. | 66 | 65 |
| 5 | IPM Demonstration under ISOPOM Pulses. | 25 | 24 |

This IPM approach will be continued during 2005-06 also.

## Pesticides Testing Laboratories

Nine Pesticides Testing Laboratories are functioning in the State with the capacity to analyse 13860 samples per annum.

These notified laboratories are functioning to enforce Insecticide Act and rules. During 2004-2005, 12721 Nos. of samples were analyzed up to February 2005. It is programmed to analyze 13860 Nos. of Pesticide samples during 2005-2006.

## CROP ORIENTED STATE SCHEMES

## Pulses Village Scheme

India is the largest producer and consumer of pulses with 33% of the World"s area and 22% of the production. In India, Tamilnadu contributes 3.2% of the total area and 2.42% of production of pulses.

There is big gap exists between demand and production of pulses. The average productivity of pulses for the past 5 years, is around 400 kgs./Ha.

The major reason for stagnation in productivity is that more than 90% of the pulses area is under rainfed condition and raise in marginal soils. Hence, a Pulses Village scheme is under implementation since 2003-04 with the objective to enhance the pulses productivity from 450 kgs./ha. to 750 kgs./ha.

This scheme is under implementation during 2004-05, in the 20 districts of Thiruvannamalai, Pudukkottai, Trichy, Perambalur, Karur, Thanjavur, Thiruvarur, Nagapattinam, Salem, Namakkal, Dharmapuri, Dindigul, Theni, Virudhunagar, Thirunelveli, Thoothukudi, Cuddalore, Erode, Vellore and Coimbatore where pulses are grown extensively to cover 280 villages at the rate of 14 villages in each district.

In each village 100 acres (40 Ha.) are being covered under Pulses totaling to 28000 Acres (11200 Ha.)

This will serve as a compact block demonstration for the neighbouring farmers to know and adopt the latest technologies.

In these demonstration plots, pulses seeds are supplied at 50% subsidy to the maximum of Rs.15/kg. besides the supply of DAP at 4 kgs. per Ac. with a maximum subsidy of Rs.20/Ac. supply of Dhal Processing Units to the maximum subsidy of Rs.10,000 each is also extended .

This scheme is under implementation at a total cost of Rs.41.20 lakhs.

The component wise physical and financial target during 2004-05 and achievements upto February, 2005 are as follows:

| Sl. No. | Component | Unit | Physical | | Financial (L.Rs.) | |
|---|---|---|---|---|---|---|
| | | | Target | Achmt. upto Feb.05 | Target | Achmt. upto Feb.05 |
| 1. | Compact Block Demo | (Ha.) | 11200 | 8560 | -- | ... |
| 2 | Seed Distribution | (MT) | 224 | 172 | 33.60 | 29.941 |
| 3 | D.A.P. Spraying | (MT) | 112 | 51.62 | 5.60 | 3.90 |
| 4 | Dhall Processor | Nos. | 20 | 10 | 2.00 | 1.00 |
| | Total | | | | 41.20 | 34.84 |

## COCONUT DEVELOPMENT PROGRAMME

In Tamil Nadu, Tall, TxD and DxT seedlings are produced in 22 coconut nurseries with two mother palm gardens to make available the required number of coconut seedlings to the farmers.

The seedlings are sold to the farmers at Rs.12/- per Tall seedling Rs.20/- per Tall x Dwarf seedling and Rs.75/- per Dwarf x Tall seedling.

During 2004-05 the production of Dwarf x Tall Hybrids is less due to limited availability to Mother Palms in Government Farms. The progress made under procurement of seed nuts and distribution of coconut seedlings during 2004-05 and programme for 2005-06 are given below: (Unit: L.Nos.)

| Sl. No. | Variety | Target 04-05 | | Achievement 04-05 | | Programme 05-06 | |
|---|---|---|---|---|---|---|---|
| | | Nuts procurement | Seedlings Distribution | Nuts Procurement | Seedlings Distribution (Feb. 05) | Nuts Procurement | Seedlings Distribution |
| 1. | Tall | 3.0 | 2.25 | 4.271 | 2.445 | 3.00 | 2.25 |
| 2. | T x D | 3.20 | 2.24 | 2.592 | 2.143 | 3.20 | 2.24 |
| 3. | D x T | 1.00 | 0.35 | 0.092 | 0.016 | 0.200 | 0.070 |
| 4. | ** | 5.30 | 4.00 | .. | .. | 7.80 | 6.00 |

** Free Seedlings distribution to female infants

The Programme will be continued during the year 2005-06 also.

## COTTON DEVELOPMENT SCHEME

The area under cotton which was around 2.67 L.Ha. during 1992-93 has declined to around one L.Ha. during 2003-04. The decline has been due to failure of monsoon and switch over by farmers to other less risky and more profitable crops like maize.

The mills require a large quantity of cotton at cheaper rate. At present, they incur heavy expenditure on transport from neighbouring states. The farmers therefore need to be motivated to increase the area under cotton and to adopt the latest technologies.

Much thrust is being laid to increase the area of cotton through sustained extension efforts backed with supply of quality inputs. There has been possibility of increasing area under irrigated and rice fallow cotton, during current year.

The rainfall received during June-December coincides with rainfed as well as winter irrigated seasons, which provides avenues for expansion of area under winter cotton. Steps are being taken to increase the area by motivating farmers to go for short duration varieties which is a remunerative alternative crop for rice fallow pulses. To encourage cotton growers in Tamilnadu contract farming is encouraged with the buy back arrangements for their produce which is detailed in a separate chapter. In order to increase area, production and productivity of cotton in Tamil Nadu, the Department of Agriculture is adopting the following strategies and schemes.

- Cultivation of hybrids
- Field demonstration on IPM and frontline demonstration on machineries and implements
- Micro Irrigation systems to maximise water use efficiency.
- Farmers Training on Production Technologies.
- Constant monitoring of pest and diseases.
- Promoting Inter cropping / mixed cropping/ Rice fallow cotton.
- Contract farming among the farmers with reputed mills.
- Insurance cover / crop loans.
- Involving Self Help Groups of farmers for area expansion.

The progress made under this scheme upto February, 2005 and the programme for 2005-06 are given below:

| Sl. No. | Particulars | 2004-05 | | Programme for 2005-06 |
|---|---|---|---|---|
| | | Target | Achievement upto Feb. 2005 | |
| 1. | Area covered under improved varieties/hybrids (L.Ha.) | 2.50 | 1.630 | 2.50 |
| 2. | Lint production (L.Bales) | 5.62 | 2.809 | 5.85 |
| 3. | Certified seed production (Qtl.) | 3670 | 1250 | 3670 |
| 4. | Certified seed distribution (Qtl) | 3670 | 1370 | 3670 |
| 5. | Seed procurement (financial: L.Rs.) | 93.65 | 61.48 | |
| 6. | NPV Distribution (Ha.) | 1500 | 1300 | 1500 |
| 7. | NPV Distribution (financial: L.Rs.) | 0.60 | 0.532 | 0.60 |

This scheme will be continued during 2005-06 also.

## STATE SCHEME ON PEST SURVEILLANCE IN COTTON

The scheme is implemented with the objective to monitor the incidence of pest and diseases, in the cotton growing areas in the 10 selected districts, through scouting thereby reducing crop losses by adopting IPM measures.

The village unemployed youth are trained in pest surveillance activities and they act as scouts for monitoring the pest incidence at village level. A scout has to cover 100 ha. and an assistance of Rs.500 is given for scout per month and totally Rs.3000/scout is given for the crop duration.

The scheme was announced by the Hon'ble Minister for Agriculture during 2002-03, and from 2002-03 onwards it is being implemented in the State. During 2004-05, an amount of Rs.1.50 lakhs has been sanctioned to cover 5000 ha. under the scheme.

| Sl. No. | Component | Target | Achmt. upto February 2005 | Programme for05-06 |
|---|---|---|---|---|
| 1. | Physical | 5000 Ha. | 3000 | 5000 Ha. |
| 2. | Financial | 1.50 L.Rs. | 0.80 | 1.50 L.Rs. |

### Tripartite Contract Farming

During 2004-05, the Department of Agriculture has organized a 'Cotton Stake Holders Meet to promote contract farming' at Coimbatore on 20.11.04 involving officials from GOI, mills, cotton farmers, CCI, Credit Institutions, Insurance Agencies, Marketing Sector, Inputs Suppliers, TNAU and Central Institute for Cotton Research. The Commissionerate of Agriculture has formulated a tripartite contract farming model for enhancing cotton production in the state and approached various agencies viz., Mills, Cotton Corporation of India, NAFED, Insurance Agencies, Banking Institutions for participation in the Programme. The main objectives of the model developed by the Commissionerate of Agriculture are as follows:

- To focus on cluster of villages.
- De-risk the farmers
- Viable price mechanism
- Credit linked inputs supply

- Insurance Coverage
- Ensure quality as per mills specifications
- Extension support by the Department

Under this model, the Department of Agriculture shall act as the Chief Co-ordinator of the Programme and TNAU to render the R&D support. Dispute settlement, if any, shall be looked after by CICR.

Under this programme, Royal Classic Mills, Palani have consented to take up contract farming in cotton during summer cotton season 2005, in the districts of Salem, Namkkal, Dindigul and Erode in an area of 5000 acres projecting the production of 50,000 Qtls and agreed to procure cotton at Rs.1850/- per qtl. or on the prevailing market rates whichever is high. The Cotton Corporation of India will act as the alternate procurement agency if the Mill fail to procure the produce. Rasi Seeds (P) Ltd., Salem, consented to supply RCH 2Bt. Seeds to the farmers covered under contract farming Programme. Royal Classic Mills so far distributed 1300 nos. of seed packets to identified 1300 nos. of farmers and imparting training to field functionaries of the districts is under progress. The Royal Classic Mills propose to cover about 20,000 Acres in 6 districts during 2005-06. The Super Spinning Mills now came forward and willing to join the Tripartite model of contract farming by covering 4,900 acres in Madurai, Salem, Erode and Theni districts during Summer 2005. The Super Spinning Mills proposed to cover about 20,000 acres during 2005-06 in 10 districts.

Our State model has been accepted by the GOI with laurels. Not only praising the efforts taken by our State, the GOI has circulated this Tripartite model of Contract Farming to all the States for emulation and also suggested that such an arrangement may augment the tradition of contract farming of cotton so that the cotton of desired quality is available to mills and also indicated that the memorandum of agreement may be helpful to State extension machinery as well as to the organization aspiring to popularize the contract farming in cotton.The Commissionerate of Agriculture will provide extension support as well as training to the farmers. Awareness campaigns will be conducted at village level. Commercial Banks will provide credit facilities to the identified farmers with Insurance coverage.

## CONTRACT FARMING IN COTTON

The cotton is cultivated in Tamil Nadu in an area of 2 lakh ha. in the normal years and the state production is around 4 - 5 lakh bales. More than 60% of the textile mills in the country are located in Tamil Nadu, which annually consumes nearly 50 lakh bales of cotton. The state production meets only 10% of the requirement. This huge gap between demand and supply has forced the mills to get their requirement from other states.

Since, there is a huge demand for cotton within Tamil Nadu and to link farmers with the assured marketing and to protect the interest of both the farmer and the industry, contract farming is promoted by the State Government through Department of Agriculture.

The farmer is provided support in diverse areas such as marketing, input, technical assistance, credit, production management, insurance coverage etc., under contract farming.

Contract Farming strategy has been introduced by private firms M/s. APPACHI Cotton Company of Pollachi and Super Spinning Mills Ltd. Coimbatore.

The firms have arranged input distribution, insurance benefits, crop loan, with buy-back guarantee to the farmers covered under the Programme. State Government has ordered waiver of 1% market cess for M/s APPACHI Cotton Company for three years from 2002-03 and proposal for waiver of market cess to the Super Spinning Mills is under consideration of the Government.

The progress made sofar is as follows:-

| Appachi CARE Foundation, Pollachi | | | | |
|---|---|---|---|---|
| Year | Area covered (Ac.) | Varieties cultivated | Farmers benefited (Nos.) | Districts |
| 2002-03 | 916 | Bunny, RCH 2, Supriya | 423 | Coimbatore, Theni, Namakkal. |
| 2003-04 | 710 | Bunny, Surabhi | 368 | Coimbatore, Namakkal, Ramnad |
| 2004-05 | 753 | Bunny, SVPR3, MCU 7, Spic 22, RCH 3600, MCU 7. | 749 | Tiruvarur, Thanjavur, Nagapattinam. |
| Total | 2379 | | 1540 | |

| Super Spinning Mills, Coimbatore | | | | |
|---|---|---|---|---|
| Year | Area covered (Ac.) | Varieties cultivated | Farmers benefited (Nos.) | Districts |
| 2002-03 | 1281 | Sara 1, Sara 2, MCU 5. | 820 | Coimbatore, Salem. |
| 2003-04 | 2558 | MCU 5, Sara 2, TCHB 213, Bunny. | 1310 | Salem, Coimbatore, Vellore. |
| 2004-05 | 2208 | Surabhi, MCU 5, MCU 7, Spic 22, Bunny. | 566 | Namakkal, Salem, Theni, Madurai, Tiruvarur, Nagai. |
| Total | 6047 | | 2696 | |
| Grand Total | 8426 | | 4236 | |

The Programme for 2005-06 is as follows:-

| Name of Mills | Proposed area coverage (acres) |
|---|---|
| Appachi Cotton company | 2500 |
| Super Spinning Mills | 20000 |
| Royal Classic Mills | 20000 |

## Contract farming through NAFED for Maize, Pulses and Oilseeds

The Hon'ble Chief Minister has announced on 5.6.2003 for adoption of alternative cropping strategy and crop diversification for delta areas wherein raising less water intensive crops were focused. The Department is taking all possible efforts to bring substantial area under less water intensive crop. The success of the alternative cropping strategy will largely depend on involvement of more number of farmers to raise less water intensive crop. The Commissionerate of Agriculture has introduced contract farming approach for maize, oilseeds and pulses with the following objectives

1. Area expansion under less water intensive crops
2. Increase in cropping intensity
3. Farmers income augmentation
4. Reduce middle men role in marketing
5. Increase in production and productivity

Under this programme NAFED will purchase the produce through buy back arrangements at prevailing market price or pre announced price whichever is higher. The Department of Agriculture will motivate and identify the farmers for this programme besides providing technical knowledge to increase the production and supply of inputs to the extent possible.

The Contract Farming Programme is proposed to be implemented as follows: (Unit in MT)

| Crop | District | Quantity proposed to be procured |
|---|---|---|
| Maize | Theni, Dindigul | 3000 |
| Groundnut | Cuddalore, Vellore, Tiruvannamalai, Villupuram, Salem | 2000 |
| Gingelly | Delta Districts | 500 |
| Pulses | Delta Districts of Thanjavur, Thiruvarur and Nagapattinam | 5000 |

This programme is commenced from January, 2005 and will be implemented till June, 2005 in about 500 villages during irrigated season. About 30,000 farmers are expected to be involved under this programme. This profitable marketing arrangement is expected to attract more farmers which will achieve the dual benefit of ensuring area expansion under less water intensive crop and to fetch good price to the farmers.

## Programme for Jatropha, Sweet Sorghum and Sugarbeet Cultivation

The Hon'ble Chief Minister of Tamil Nadu has announced a massive scheme for cultivation of Jatropha, Sweet Sorghum and sugar beet as a part of Alternative Cropping Strategy on contract farming basis for the production of ethanol and bio diesel. This scheme is proposed to be taken up linking Industrial entrepreneurs with buy back arrangements at pre-fixed price for the produce. These crops are less water intensive and highly remunerative to the farmers.

## Promotion of Jatropha

The Promotion of Jatropha cultivation by contract farming is entrusted to 4 entrepreneurs identified by the Government by allotting specific district of jurisdiction to them as follows.

| Firm | District allotted | Action Programme (Ha.) | |
|---|---|---|---|
| | | 2004-05 | 2005-06 |
| M/s. Mohan Breweries and Distilleries Limited | Thiruvallur | 5000 | 80000 |
| M/s. Shiva Distilleries Limited | Karur, Dindigul, Erode and Coimbatore | 2100 | 3800 |
| M/s. Dharani Sugars and Chemicals Limited | Virudhunagar | Firms yet to make their commitment | |
| M/s. Riverway Agro Products Private Ltd. | Tirunelveli and Thoothukudi | | |

The extent of Jatropha cultivation organized by M/s Mohan Breweries and Distilleries Limited is 119 Ha. and M/s. Shiva Distilleries Limited is 180 Ha. upto February 2005, as against the programme to cover 5000 Ha. and 2100 Ha. respectively. Due to non-availability of quality seed material, the area expansion gets delayed. The firms have taken steps to obtain quality seeds from Chattishgarh and Rajasthan and Nationalised banks were being persuaded by them to get credit facility. The preparations of Bipartite and Tripartite agreements is in progress to enter into the Contract Farming. The procurement rate is committed as Rs.5000/- per tonne. M/s. Mohan Breweries Limited have proposed to establish an extraction cum transesterification unit. The Government have announced that there will be no purchase tax on Jatropha. Sales tax on bio diesel from Jatropha seeds will be 4% besides issue of special incentives as per industrial Policy statement 2003 of the Government of Tamil Nadu.

# 5

# Sustainable Agriculture, Globalisation and Biotechnology

## AGRICULTURAL PROTECTION AND SUPPORT

A major impediment to growth in agricultural trade remains to be the high level of protection and support to agriculture in the developed countries. The Uruguay Round initiated the process of opening up new opportunities for export diversification in agriculture, through, *inter alia*; tariffication and across-the-board reductions in most-favoured nation (MFN) tariffs on agricultural products; reduction of trade-distorting support; and the strengthening of trade rules, particularly those on sanitary and phytosanitary measures and technical barriers to trade. Despite these changes, little improvement has been achieved and market access problems persist particularly for processed agricultural products.

*Agricultural Tariffs:* Agricultural tariffs remain high and complex. Tariffs in the developed countries are generally lower on average, but there are many tariff peaks, particularly on processed products and temperate-zone products such as horticulture, sugar, cereals, dairy products and meat. These tariffs are often complex with several different *ad valorem*, specific and seasonal rates applying to the same products. Bound tariffs are also high in many developing countries, although their applied rates are generally lower. Recent estimates show that nominal protection rates in the OECD countries have declined somewhat but remain high on the whole. While tariff rate quotas (TRQs) have created some new trading opportunities, consistently low and declining fill rates indicate that there are difficulties in their implementation.

*Tariff Escalation:* The existence of tariff escalation (tariffs which escalated with the degree of processing) in agricultural markets is regarded as one of the major factors that hinder export growth and diversification in the exporting countries. Tariff escalation prevails in a large number of agricultural commodity chains. Of 17 commodity chains examined by the Secretariat, 12 suffer from tariff escalation, mostly at the first stage of processing. In developed countries, escalation is particularly pronounced in coffee, cocoa, oilseeds, fruit and hides and skins. In developing countries, however, applied tariffs are often much lower than bound rates and in almost all cases they also show relatively high escalation.

Tariff escalation has been raised as one of the important market access issues in the current WTO negotiations on agriculture. Country proposals addressing tariff escalation are suggesting the adoption of a harmonising reduction formula that reduces higher tariffs by greater amounts, including tariff peaks, and eliminates tariff escalation. A recent FAO study shows that while the adoption of a harmonising formula for reducing tariffs such as the Swiss formula appears to be a better option for reducing tariff escalation than linear methods, it is unlikely that this formula will achieve substantial reduction in tariff escalation in all commodities of export interest to low income agricultural commodity-dependent exporters, particularly tropical products, unless it involves some additional criteria for capping tariff wedges along the processing chain.

***Support to Production and Exports:*** Subsidized production and exports affect both domestic and international markets and exert a negative influence on the diversification of production and exports from commodity exporting countries. Loss of competitiveness caused by subsidized agriculture discourages investments in agriculture and local processing in non-subsidizing countries. Under the WTO Agreement on Agriculture (AoA), WTO members undertook to reduce both export subsidies and the trade-distorting domestic support. Export subsidies have been reduced somewhat on several products since 1995, but they remain high, particularly for meat and dairy products. In domestic support, some WTO members have shifted to less distorting support measures, but overall support to agriculture remains high. Estimates show that total transfers to agriculture in the OECD countries amounted to US$311 billion in 2001, compared with US$298 billion in 1986-88. The support to producers in these countries as a share of the value of farm output averaged 31 percent for agriculture as a whole and exceeded 40 percent in such products as rice, wheat, sugar, beef and sheep meat. Such support is believed to have stimulated overproduction creating a downward pressure on world prices, thus limiting trading opportunities for non-subsidizing exporters.

***International Technical Standards***. A major challenge faced by many exporters of agricultural products, particularly in low income countries, is to meet the sanitary, phytosanitary and technical requirements of importing countries. An additional challenge is faced by these countries when new standards are introduced on risk assessment grounds that are stricter than those currently recognized by international standard-setting bodies. Rising consumer concerns in the affluent countries over food quality compound the difficulty of these countries in meeting ever higher standards. The transition to risk-based standard setting has required major changes in legislative, regulatory and administrative practices in most countries all of which have implied significant costs.

The concerns of low income countries in this regard are three-fold: first that they usually lack the technical and financial capacity to fully participate in activities of international standard-setting bodies such as the Codex Alimentarius Commission (CAC) and the International Plant Protection Convention (IPPC). Second, they lack the supply-side capacity to meet the increasingly strict standards being adopted by the developed countries. Third, there are concerns about the scope for the discriminatory use of these measures. A recent FAO study provided a broad indication of this problem based on data on rejections following border inspection in export markets. The import detention list for the United States for the period 1996-97 shows that the majority of detentions and rejections of products coming from developing countries, particularly those of Sub-Saharan Africa, are related to microbiological contaminants and filth rather than to strictly technical considerations. This illustrates the considerable problems that low income countries have in meeting basic food hygiene requirements let alone requirements for which more sophisticated monitoring and testing, and therefore more costly procedures are required, such as limits on pesticide residues and heavy metals.

## PREFERENTIAL MARKET ACCESS

Despite the current constraints on agricultural trade emanating from protection and support in the developed country markets, preferential trade arrangements involving developing countries have provided some opportunities for these countries to expand and diversify their agricultural exports. Three major forms of trade preferences can be distinguished: the Generalized System of Preferences (GSP), special preferential regimes for groups of developing countries (such as Lome/Cotonou, AGOA, or the Caribbean Basin Initiative), and regional free-trade areas between developed and developing countries. However, the last form differs from the others in that it involves reciprocal rather than unilateral trade preferences. MFN duties are still high for many agricultural products and preference schemes are offering substantial price advantages to some exporters for numerous eligible products. This is specially the case, on a wide scale, for least developed countries, as well as under special preferential arrangements of the EU and the United States for most African, Caribbean, Central American, Andean and Pacific countries.

While many developing countries have been accorded preferences for a large number of products, exceptions to these preferences often relate to agricultural products. For example, the European Union's initiative on "Everything but Arms" (EBA) offers free market access to LDC products, with less than 5 percent of pre-EBA exports left facing a tariff barrier. A recent study by UNCTAD, however, shows that the impact of this initiative will be a relatively small increase, in the medium term, in exports from the LDCs, as 70 percent of the potential positive trade effects would have come from free access for sugar, rice and beef, which has been deferred until 2006. The African Growth and Opportunity Act has also started to open up markets in the United States for African countries. Although trade preferences are of significant benefits to a number of countries, in general they have been, to a large extent, underutilised. Part of the reason for this underutilisation is due to, in many cases, lack of security of access, insufficient product coverage, excessively stringent rules of origin and often a lack of awareness of the preferences available. These factors are in addition to supply-side constraints faced by some countries, in particular the least developed countries

## DISTRIBUTION OF GAINS

Government barriers to trade are important, but as they are reduced other factors impeding market entry, particularly those that spring from restrictive business practices, come to the forefront.

The agricultural commodity chains, particularly those of high value crops and processed products, are increasingly dominated by a few multinational enterprises (MNEs) and distribution companies. This dominance of large companies in the world agricultural commodity markets is not only through their actions on international markets, but also through their direct and increasing influence on what is produced, and how. While these changes are opening up unprecedented opportunities for some producers and exporters, benefiting from these developments and avoiding their negative impacts requires much greater business skills than before on the part of Governments and small-scale entrepreneurs, particularly in low-income countries.

Changes in agricultural commodity chains, as a result of the increasing dominance of large companies, can be reviewed at three levels, namely, the international markets, exporting developing countries, and the retail markets of importing countries. At the international level, agricultural trade continues to be concentrated in a few vertically integrated MNEs. In 1996, for example, five MNEs accounted for about 50 percent of world trade in green coffee, up from about 37 percent in 1980; and four coffee roasters accounted for 50 percent of the roasted coffee market. In cocoa, the number of trading houses in London has been reduced from 30 in 1980 to around 10 in 1999. Similarly, the six largest chocolate manufacturers account for 50 percent of world chocolate sales. For vegetable oils, following a series of mergers and acquisitions in the 1990s, a small number of vertically integrated MNEs now dominate the production, distribution and international trade of both oilseeds and oils. In grains, a few big companies are transforming themselves, through mergers and consolidations, into vertically integrated businesses in trading, storage, processing and milling. In exporting developing countries, following the deregulation and the disappearance of many marketing boards, large companies with warehousing and shipping facilities in the producing countries are able to exploit their financial and logistical advantages, even buying the produce directly from the farmer. Intensified competition favours those with access to cheaper finance and good logistics. Being big provides advantages on both accounts.

At the retail level, supermarkets have also being growing rapidly in both developed and developing countries. For example, there has been rapid consolidation and multinationalisation of the supermarket sector in Latin America, where the share of supermarkets in food retailing went from 15-20 percent in 1990 to 60 percent in 2000. Given their dominance in the market, supermarkets have a significant impact on production, distribution and trade. To simplify operations, most of these supermarkets are increasingly cutting the number of their suppliers, and as a result those left out will have to sell in a shrinking and low-priced market or find a niche market. In addition, supermarkets have significant impact on production requirements in terms of quality and food safety, product variety and delivery. The Food Business Forum (CIES), through its association of the world's 200 largest supermarkets and their suppliers (with a combined sales of US$2 800 billion), decided earlier this year on new mandatory standards for their suppliers. While this move may help in reducing the number of different standards, it may overshadow whatever Governments decide in the context of the WTO agreements.

The stringent private standards set by supermarkets are very often a reflection of consumers' concerns. Consumers of agricultural commodities are increasingly choosy with respect to the quality of the commodities they buy, the timeliness of delivery, origin and traceability, and the environmental and social conditions under which they have been produced. The use of specialised channels such as fair-trade labelling and other socially or environmentally responsible market practices, and taking advantage of niche market such as that for organic food, may help in facilitating market entry and can improve earnings of producers. In many countries, consumers are willing to pay a price premium (of more than 20 percent) for organic foods, particularly fruit and vegetables.

There is evidence that the increasing market concentration along agricultural commodity chains has been associated with a declining share of producing countries in the revenues generated from the sale of these products in global markets. Recent studies show a widening gap between world commodity prices and consumer prices in industrialized countries. Growers' price represents very low shares of the final price of the product, ranging from 4-8 percent for raw cotton and tobacco to 11– 24 percent for jute and coffee. Similar patterns were also observed in fresh fruit and vegetables. The International Coffee Organisation (ICO) has reported that in the early 1990s export earnings by coffee producing countries were about US$10-12 billion and the value of retail sales of coffee, largely in the developed countries, was about US$30 billion. However, in coffee year 2000/01, producing countries received only US$5.8 billion of the value of retail sales of more than US$60 billion.

Studies on commodity chains involving exports of fresh vegetables and fruit from Africa to the UK have shown that returns to the participant in the market are highly concentrated at the end of the chain in the importing countries. As shown in the figures below, returns to producer, exporter and packaging of fresh vegetables exported from Kenya and mangetout from Zimbabwe amounted to only 27 and 23 percent of the final price of these products respectively.

## FOCUS ON ISSUES FACING FARMERS AND PRODUCERS

The issues that are important to farmers across the country include stewardship, profitability, sustainability and the health of their community. And those are the topics to be covered in Kansas in an upcoming agriculture roundup.

The practice of sustainable agriculture is built upon soil fertility and the protection of soil health. For centuries, farmers around the world have been employing these basic techniques to keep their soil productive. Early in colonial American history, George Washington was using cover crops and manure applications on his farm. American farmers, who did not carefully tend their soil, eventually "wore it out." The Westward Expansion partly reflected the lowered productivity of eastern lands and the search for new farmland farther west. The first publications from the Kansas State Agricultural College appeared in the late 1880's, at a time when Kansas soils had been tilled for less than fifty years. Soil scientists were well aware at that time that the long term sustainability of farming depended upon the use of legumes for fertility and the addition of organic matter for tilth. As Kansas farmers began to report declining fertility levels in the early 1900's, scientists cautioned that farmers must not rely solely upon the natural fertility of the prairie soils.

Between 1900 and 1950, Kansas State researchers guided farmers in the use of legumes and the preservation of soil organic matter. With the advent of inexpensive nitrogen fertilizers after World War II, traditional soil management techniques took a back seat to the use of commercial products that were easy to use and often provided greater economic returns in the short-term. At the beginning of the twenty-first century, farmers are becoming increasingly conscious of the importance of soil health, water quality, and energy conservation. The rising cost of nitrogen fertilizers has revived interest in nitrogen-fixing legumes. Excess phosphorus levels in surface water indicate a need to emphasize soil conservation and the careful use of manure resources. Traditional soil management practices continue to be vital for the sustainable agriculturist and are regaining an audience with conventional agriculture. This review of Kansas State University soil publications profiles the research and recommendations of Kansas scientists during the early to mid-twentieth century. Generally, recommended practices such as crop rotations, manure use, cover crops, and other sources of fertility are considered including the shift to commercial fertilizers in the 1950's. The special consideration of western Kansas soils is treated separately. Limited rainfall in western Kansas affects the research conducted in that part of the state and alters the use of traditional soil health practices.

## SOIL HEALTH

Within every aspect of Kansas agriculture, healthy soil is a key element. Its structure and fertility provide the basis for all crop and livestock production. Throughout the historical publications of Kansas State University from the late 1800's to the mid-1900's, researchers and educators have been concerned with the protection of this vital resource. In their 1918 publication, Soil Fertility, L.E. Call and R. I Throckmorton caution the Kansas farmer. "The soil is the most important source of wealth in an agricultural state. If it is maintained in a high state of productivity, by wise systems of soil management, the people prosper. If its fertility is wasted through careless methods of farming, both the farmer and the state suffer". Forty years later, in 1956, Orville Bidwell echoes this same message with its promise of a precarious wealth. "Unlike most other resources, soil is inexhaustible if properly managed." Bidwell recounts the variety of Kansas soils each with a different waterholding capacity, permeability, response to fertilizers, and susceptibility to erosion.

## DECLINE OF KANSAS SOILS

Throughout many of the earliest publications, the authors are clearly concerned about the declining condition of Kansas soils. By the early 1900's, much of the rich prairie soil in eastern Kansas had been farmed for 50 years. A 1903 publication from the veterinary department of the Kansas State Experiment Station states, "The fertility of the soil of the Middle states and the West is being rapidly diminished and if means are not taken to prevent it, the time is not far distant when it will be as necessary to apply artificial fertilizers to the soil as it is now in the East".

A few years later, chemists at the agricultural experiment station raised the same concern. "In the early history of Kansas no attention was paid to the composition of its soils except to boast of their inexhaustible fertility. The voice of the chemist has been lifted constantly, warning the people that this idea of possession of a fertility that is practically limitless is a delusion that can lead only to squandering of our natural resources, and to leaving posterity handicapped in the struggle for existence. Today he is seeing his warnings justified. People in many localities of the eastern part of the state are making inquiry concerning chemical analysis of their soils with reference to learning what fertilizers should be applied and to what crops their soils are best adapted".

By 1918, L.E. Call and R. I. Throckmorton attempted to put dollar figures to the losses in soil fertility. They estimated that the plant food removed from Kansas soils by wheat crops over the previous fifty-five years equalled a value of more than seven hundred million dollars. Even the wheat straw, which was regularly burned or wasted, had a value in plant nutrients of more than twelve million dollars. The majority of the wheat products were both milled and eaten outside the state which Call and Throckmorton equated with the export of soil fertility. Call and Throckmorton credited the declining productivity of Kansas soils to five factors: depletion of soil organic matter, failure to grow enough acres of leguminous crops for nitrogen fixation, depletion of mineral nutrients, the lack of proper crop rotations, and the erosion of fertile topsoil. These five factors are the subject of nearly every soils publication prior to the advent of inexpensive nitrogen fertilizers in the mid-1900's. They continue to be the basis of soil health for every farming system regardless of the use of commercial fertilizers.

## Soil Organic Matter

The early Kansas State publications recognized the importance of soil organic matter to soil health. Replenishment of soil organic matter is the most basic step in addressing a number of other issues. According to Call and Throckmorton, organic matter holds the soil's store of nitrogen, provides good tilth, holds moisture, and provides food for the bacteria that make nutrients available to plants.

Soil studies in western Kansas spanning over 30 years of crop production give mention to the importance of organic matter. Progress reports in 1943 and 1957 indicate that as the organic carbon content of the soil has decreased, more power is needed for tillage, water intake is decreased, seedling emergence and root growth are hindered. The 1943 report also mentions that the presence of coarse, fibrous organic matter helped reduce wind erosion. Numerous publications cite frequent plowing and intensive cultivation as culprits in the loss of organic matter. Row crops depleted the soil more quickly than small grain crops. Without a regular practice of restoring organic matter, the levels of carbon in the soil would drop, threatening fertility and soil structure.

The 1918 publication, Soil Fertility, promotes barnyard manure applied to the soil as a primary source of organic matter. However, when sufficient manure is not available, the farmer should grow a crop to plow under to supply organic matter. These green manure crops may be legumes such as alfalfa, cowpeas, soybeans, clover, or sweet clover. Legumes would capture atmospheric nitrogen and fix it in the soil in addition to supplying organic matter. Non-legume crops such as rye, buckwheat, sorghum, and turnips were also cited as possible sources for organic matter.

Although most of the methods for increasing organic matter focus on the addition of green manure crops or barnyard manure, the conservation of all organic matter is addressed. Throckmorton and Call state that most wheat straw is either burned or destroyed after threshing. They recommend its use as feed and bedding for livestock. Eventually the soiled bedding and any manure would be returned to the soil. They also suggest use of the straw as a surface mulch on wheat during the winter at a rate of 1-1.5 tons per acre. In 1962, farmers were again cautioned not to waste the organic matter provided by straw or stubble. In an effort to control weeds and increase yields, burning of wheat stubble had become a common practice. A series of studies in western

Kansas produced data showing that stubble burning did not increase yields for subsequent crops. Not only did it present increased potential for wind erosion, it also decreased the soil's ability to absorb water.

## NITROGEN AND ORGANIC MATTER

The supply of nitrogen in the soil was closely linked to the organic matter content. The practice of returning organic matter to the soil in the form of leguminous plants or animal manures was also the method for supplying nitrogen prior to the use of commercial nitrogen fertilizer. Legume crops, grown in rotation with other crops, could be worked back into the soil as a green manure crops or the legume hay crop was fed to livestock and their manure was returned to the soil.

Off-farm sources of nitrogen in the early 1900's included waste materials from the packing houses and inorganic compounds such as saltpeter, which was mined in Chili, and manufactured compounds of ammonium sulphate or calcium cyanamide. Researchers noted that "the purchase of nitrogenous fertilizers should be limited to the meeting of special requirements of certain conditions or crops...Nitrogen...is the (nutrient) most cheaply restored, since by the cooperation of clovers, alfalfa, peas, beans, and other legumes with bacteria that grow upon their roots the abundant nitrogen of the air in the pores of the soil is brought into organic combination. This means of adding nitrogen to a soil must never be lost to view...".

Through the first half of the twentieth century, researchers and farmers sought to understand the best methods for capturing nitrogen with legumes especially when those crops were being used for other purposes on the farm. In 1918, researchers were concerned that the nitrogen found within the crop roots and stubble might not be significant compared with the amount removed in a hay crop. They hypothesized that a certain amount of leaf loss during the haying process might be returning some nitrogen to the soil. Still they cautioned that the best practice was to feed the hay on the farm and return all manure to the soil. The 1939 publication on fertility studies at the Manhattan experiment station beginning in 1910 gave evidence that soil nitrogen could be increased even when the hay was removed. The average increase in the 5-9 year old hay plots was twice that for the 1-4 year plots. The researchers also found that the residual effects of the "nitrogen accumulating capacity" remained for eight to nine years following the breaking of the alfalfa sod that had been in alfalfa for more than two years.

But nitrogen was not the only concern when hay was taken off the farm. "If alfalfa is sold off the land it is one of the most soil-exhausting crops raised, while if it is fed on the farm, and the manure produced applied to the land, it is a conserver of fertility. The same argument applies to clover". Potassium, phosphorus, calcium and other mineral elements are taken up by the plants and must be cycled back to the soil as green manure or barnyard manure. When hay is sold off the farm, the phosphorus and potassium are more rapidly depleted than with a continuous grain crop.

Regardless of the use of green manure crops, any farm with livestock had a source of organic matter and nutrients in animal manure if it was used wisely. "If the livestock farmer properly saves and utilizes his manure he can maintain his soil in a high state of productivity, but the livestockman who feeds his cattle in woodlots along the banks of streams, and so wastes his manure, usually depletes the fertility of his soil more rapidly than the man producing grain only.

It was important to manage manure so that nutrients were not lost before they were cycled back to the soil. The seepage of liquid waste or urine, leaching of manure by rain and runoff water, and the "decay" of manure solids and the subsequent loss of nitrogen are all a result of poor handling techniques. Throckmorton and Call recommend that the most practical method of manure handling would be to feed the stock on the cultivated fields so that the manure is scattered by the animals and the nutrients are retained by the soil. In any case, the manure should be returned to the soil as soon as possible and long term, open storage of six months or more should be avoided. Manure could also be used sparingly as a top dressing on corn, kafir, or winter wheat where it would act as mulch to retain moisture. If there was not a large supply of manure, it was deemed better to apply the available supply lightly to more acres rather than a heavy application on just a few acres.

As a percentage of volume, the nutrients present in manure are small. "The figures for the fertilizing constituents are always low, but they are present in readily available form and the accompanying organic matter has itself a highly beneficial effect on the land. Even with these low percentages the total amount of plant food in the manure produced on a farm reaches very significant quantities".

## Phosphorus

Fertilizers were commonly used to provide phosphorus and other minerals long before nitrogen fertilizers were in general use. By 1918, the soils in eastern and southeast Kansas had such low stores of phosphorus that it was profitable to purchase phosphorus fertilizers. Grain farms lost phosphorus the most rapidly when it was exported off the farm with the grain. In order to retain the mineral, the farmer had to feed the grain to livestock and apply the manure to the croplands or else import a supply of phosphorus.

Alfalfa and other deep-rooted crops could be used to collect phosphorus from the subsoil. The nutrients then needed to be cycled back into the upper levels of the soil as either a green manure crop or manure from animals fed on the hay.. Commercial sources of phosphorus included bone, basic slag, rock phosphate, and apatite. Bones were a valuable product rich in phosphorus and nitrogen. They could be ground raw but they were more commonly steamed before grinding. This processing made grinding easier, concentrated the phosphorus slightly and increased the availability to plants.

Basic slag was a by-product from a particular method of iron refinement. Minerals removed from certain iron ore contained high amounts of phosphate, which could be ground for use as a fertilizer. This was not a common product in America, being chiefly available in Europe. Primary sources of rock phosphate were found in Florida, South Carolina, and Tennessee during the early part of the twentieth century. Although it was most often used to produce superphosphate, it could be finely ground and used raw. Superphosphate was produced by treating the raw phosphate with sulphuric acid and was in use to some extent throughout the twentieth century.

Although superphosphate was more readily available to the plant, Kansas State researchers advised that the raw phosphate was usually a wiser choice. "In the application of phosphate fertilizers the farmer naturally expects and desires immediate results, which are secured by the use of superphosphate, but at the same time if larger quantities of phosphates can be applied in other less soluble and available forms at the same expenditure, the ultimate value of the investment may be much greater, as the phosphorus will remain in the soil and be rendered available by slow natural processes. There was also indicated a positive relationship between the bacteria found in organic matter and the availability of phosphorus from raw phosphate. "The cheapest source of phosphorus is ground rock phosphate, and in this form it will be available for the use of crops if the soil is well supplied with organic matter from farm manures and legumes. Apatite, a crystal found in granite, was abundantly available in Canada. It needed to be converted to superphosphate to be used as a fertilizer.

## Potassium

Potassium occurs in the mineral or rock portion of the soil. Plant roots are able to access potassium directly from tiny soil particles, chiefly silt and clay. Potassium is also present in decaying crop residues and organic matter, indicating the need, once again, for returning plant materials to the soil. A traditional source of additional potassium was hardwood ashes. Muriate of potash, a processed form of potassium in wide use today, was also available early in the twentieth century but due to the chlorine content of the compound, its use was restricted with certain crops.

## Calcium (Lime)

Calcium, while an essential element for plant growth, was usually not deficient in Kansas soils. However, its application in the form of lime, was a common soil amendment. "(Calcium) is generally present in all cultivated soils in sufficient quantity to supply fully the need of the plant. Yet even where this is the case, the soil may be greatly in need of liming. Lime is used, therefore, as a soil amendment not so much for its effect directly on the plant as for its effect on the soil, which indirectly affects the plant. Soils that are low in lime are said to be sour".

The acidity of these soils could be neutralized by the application of calcium compounds. A 1918 publication recommended two tons of ground limestone per acre followed by one or two tons every five to six years thereafter. The limestone should be worked into the bare ground six months to a year prior to seeding alfalfa or clovers. The effects of the limestone are slow and gradual but long lasting. The bacteria that live on the roots of alfalfa, sweet clover, and clover thrive in more alkaline soils so these crops respond well to liming. Crops considered less sensitive to acid soils were corn, wheat, timothy, and oats.

The change in soil pH improved the availability of phosphorus and potassium as well as improving the texture of the soil. "It is a well-known fact that soils well stocked with a supply of lime will be more productive under the same conditions of plant-food content than soils not so stocked...While soils use very small amounts of lime as compared with phosphorus and potassium, yet the presence of a relatively large supply of lime insures crop production. In the words of Hilgard, 'A lime country is a rich country'".

## SOIL BACTERIA

Although the exact relationship did not seem to be clear, researchers in the early part of the twentieth century were writing about a correlation between humus or organic matter, bacteria, and fertility. A series of studies beginning in the 1890's showed crop yields in "direct proportion" to the bacterial content of the fields. In this same publication, the authors speculate that fertility depends to a large extent on bacterial activity and that by manipulating the bacteria of the soil, one might avoid the need for artificial fertilization. They proposed additional experiments to find ways to increase soil bacteria. A few years later, Walter King and Charles Doryland report on their studies of soil bacteria with this same basic assumption. Assuming that bacterial activity indicated increased soil fertility, they set about collecting samples from various soil depths on plots representing a variety of tillage practices. They admitted tremendous variability in their data, which was collected over a period of only three months from March to June 1908. Nevertheless, they concluded that deep plowing, conveniently the tillage practice most commonly used at that time, increased bacteria levels and bacterial activity and decreased denitrification. They noted that bacterial activity increased with the temperature of the soil and decreased when the soil became saturated with moisture. Different species were predominate at different times and activity seemed to rise and fall with a regularity independent of moisture and temperature.

## THE TECHNICAL PROCESSING OF AGRICULTURAL PRODUCE

Above we have already had occasion to note (Chapter I, § I) that writers on agriculture, in classifying systems of farming according to the principal market product, assign the industrial or technical system of farming to a special category. The essence of this system is that the agricultural product, before going into consumption (personal or productive), undergoes technical processing. The establishments which effect this processing either constitute part of the very farms on which the raw material is produced or belong to special industrialists who buy up the raw material from the peasant farmers. From the standpoint of political economy the difference between these two types is unimportant. The growth of agricultural technical trades is extremely important as regards the development of capitalism. Firstly, this growth represents one of the forms of the development of commercial farming, and is, moreover, the form that shows most vividly the conversion of agriculture into a branch of industry of capitalist society.

Secondly, the development of the technical processing of agricultural produce is usually connected intimately with technical progress in agriculture: on the one hand, the very production of the raw material for processing often necessitates agricultural improvement (the planting of root-crops, for example); on the other hand, the waste products of the processing are frequently utilised in agriculture, thus increasing its effectiveness and restoring, at least in some measure, the equilibrium, the interdependence, between agriculture and industry, the disturbance of which constitutes one of the most profound contradictions of capitalism. We must accordingly now describe the development of technical agricultural trades in post-Reform Russia.

## *Distilling*

Here we regard distilling only from the point of view of agriculture. Accordingly, there is no need for us to dwell on the rapid concentration of distilling in large plants (partly due to excise requirements), on the rapid progress of factory technique, with the consequent cheapening of production, and the increase in excise duties which has outstripped this cheapening of production and because of its excessive amount has retarded the growth of consumption and production.

Thus, over 9/10 of the distilleries (accounting for over 4/5 of the total output) are directly connected with agriculture. Being large capitalist enterprises, these establishments lend the same character to all the landlord farms on which they are set up (the distilleries belong almost without exception to landlords, mainly to members of the nobility). The type of commercial farming under review is particularly developed in the central black-earth gubernias, in which are concentrated over 1/10 of the total number of distilleries in the Russian Empire (239 in 1896-97, of which 225 were agricultural and mixed), producing over a quarter of the total output of spirits (7,785,000 vedros in 1896-97, of which 6,828,000 at agricultural and at mixed distilleries). Thus in the area where labour-service predominates, the commercial character of agriculture most frequently (as compared with other areas) manifests itself in the distilling of vodka from grain and potatoes. Distilling from potatoes has undergone a particularly rapid development since the Reform, as may be seen from the following data relating to the whole of the Russian Empire:

Thus, with a general twofold increase in the quantity of crops distilled, the quantity of potatoes used increased about 15-fold. This fact strikingly corroborates the proposition established above (§ I in this chapter) that the enormous increase in the potato area and crop signifies the growth of precisely commercial and capitalist farming, along with improvement of agricultural technique, with the replacement of the three-field system by multi-field crop rotation, etc. The area of the biggest development of distilling is also distinguished for the biggest (in the Russian gubernias, i.e., not counting the Baltic and the western gubernias) net per-capita harvest of potatoes. Thus in the northern black-earth gubernias the figures for 1864-1866, 1870-1879 and 1883-1887 were 0.44, 0.62 and 0.60 chetverts respectively, whereas for the whole of European Russia (50 gubernias) the corresponding figures were 0.27, 0.43 and 0.44 chetverts. As far back as the beginning of the 80s the *Historico-Statistical Survey* noted that "the region marked by the greatest expansion of potato cultivation covers all the gubernias of the central and northern parts of the black-earth belt, the Volga and Transvolga gubernias and the central non-black-earth gubernias" (*loc. cit.*, p. 44). The expansion of potato cultivation by landlords and well-to-do peasants means an increase in the demand for hired labour; the cultivation of a dessiatine of potatoes absorbs much more labour than the cultivation of a dessiatine of cereals and the use of machinery in, for example, the central black-earth area is still very slight. Thus, while the number of workers engaged in the distilling industry proper has decreased, the elimination of labour-service by the capitalist system of farming, with the cultivation of root-crops, has increased the demand for rural day labourers.

### *Beet-Sugar Production*

The processing of sugar-beet is even more highly concentrated in big capitalist enterprises than distilling is, and is likewise an adjunct of the landlords' (mainly noblemen's) estates. The principal area of this industry is the south-western gubernias, and then the southern black-earth and central black-earth gubernias. The area under sugar-beet amounted in the 60s to about 100,000 dess., in the 70s to about 160,000 dess.; in 1886-1895 to 239,000 dess., in 1896-1898 to 369,000 dess., in 1900 to 478,778 dess., in 1901 to 528,076 dess. (*Torgovo-Promyshlennaya Gazeta*, 1901, No. 123), in 1905-06 to 483,272 dess. (*Vestnik Finansov*, 1906, No. 12). Hence, in the period following the Reform the area cultivated has increased more than 5-fold. Incomparably more rapid has been the growth of the amount of sugar-beet harvested and processed: on an average the weight of sugar-beet processed in the Empire in the years 1860-1864 was 4.1 million berkovets; in 1870-1874—9.3 million; in 1875-1879—12.8 million; in 1890-1894—29.3 million; and in 1895-96 and 1897-98—35 million. The amount of processed sugar-beet has grown since the 60s more than 8-fold. Hence, there has been an enormous increase in the beet yield, i.e., in labour productivity, on the big estates organised on capitalist lines. The introduction of a root-plant like beet into the rotation is indissolubly linked with the transition to a more advanced system of cultivation, with improved tillage and cattle feed, etc. "The tillage of the soil for beetroot," we read in the *Historico-Statistical Survey* (Vol. I), "which, generally speaking, is rather complicated and difficult, has been brought to a high degree of perfection on many beet farms, especially in the south-western and Vistula gubernias. In different localities, various more or less improved implements and ploughs are used for tilling; in some cases even steam ploughing has been introduced" (p. 109).

This progress of large-scale capitalist farming gives rise to quite a considerable increase in the demand for agricultural wage-workers— regular and particularly day labourers—the employment of female and child labour being particularly extensive (cf. *Historico-Statistical Survey*, II, 32). Among the peasants of the neighbouring gubernias a special type of migration has arisen, known as migration "to sugar" (*ibid.*, 42). It is estimated that the complete cultivation of a morg (= 2/3 dess.) of beet land requires 40 working days (*Hired Labour*, 72). The *Combined Material on the Position of the Rural Population* (published by Committee of Ministers) estimates that the cultivation of one dessiatine of beet land, when done by machine, requires 12, and when by hand 25, working days of males, not counting women and juveniles (pp. X-XI). Thus, the cultivation of the total beet area in Russia probably engages not less than 300,000 agricultural day labourers, men and women. But the increase in the number of dessiatines under beet is not enough to give a complete idea of the demand for hired labour, since some jobs are paid for at so much per berkovets. Here, for example, is what we read in *Reports and Investigations of Handicraft Industry in Russia* (published by Ministry of State Properties, Vol. II, St. Petersburg, 1894, p. 82).

"The female population both of the town, and of the uyezd" (the town of Krolevets, Chernigov Gubernia, is referred to) "think highly of work on the beet fields; in the autumn the cleaning of beets is paid at 10 kopeks per berkovets, and two women clean from six to ten berkovets a day, but some contract to work during the growing season as well, weeding and hoeing; in that case, for the full job, including digging and cleaning, they get 25 kopeks per berkovets of cleaned beets." The conditions of the workers on the beet plantations are extremely bad. For instance, the *Vrachebnaya Khronika Kharkovskoi Gubernii* (September 1899, quoted in *Russkiye Vedomosti*, 1899, No. 254) cites "a number of exceedingly deplorable facts about the conditions *of those working on the red-beet plantations*. Thus, the Zemstvo physician, Dr. Podolsky, of the village of Kotelva, Akhtyrka Uyezd, writes: 'In the autumn *typhus* usually breaks out among *young people employed on the red-beet plantations of the well-to-do peasants*.

The sheds assigned for the workers' leisure and sleeping quarters are kept by such planters in a very filthy condition; by the time the job ends the straw used for sleeping is literally converted into dung, for it is never changed: this becomes a breeding ground of infection. Typhus has had to be diagnosed immediately in the case of four or five patients brought in from one and the same plantation.' In the opinion of this doctor, 'most of the syphilis cases come from the red-beet plantations.' Mr. Feinberg rightly asserts that 'work on the plantations, which is no less injurious to the workers themselves and to the surrounding population than work in the factories, has particularly disastrous consequences, because large numbers of women and juveniles are engaged in it, and because the workers here are without the most elementary protection from society and the State'; in view of this, the author wholly supports the opinion expressed by Dr. Romanenko at the Seventh Congress of Doctors of Kharkov Gubernia that 'in issuing *compulsory* regulations, *consideration* must also be given *to the conditions of the workers on the beet plantations.* These workers lack the most essential things; they live for months under the open sky and eat from a common bowl.'" Thus, the growth of beet cultivation has enormously increased the demand for rural workers, converting the neighbouring peasantry into a rural proletariat. The increase in the number of rural workers has been but slightly checked by the inconsiderable drop in the number of workers engaged in the beet-sugar industry proper.

## *Potato-Starch Production*

From branches of technical production conducted exclusively on landlord farms let us pass to such as are more or less within the reach of the peasantry. These include, primarily, the processing of potatoes (partly also wheat and other cereals) into starch and treacle. Starch production has developed with particular rapidity in the post-Reform period owing to the enormous growth of the textile industry, which raises a demand for starch. The area covered by this branch of production is mainly the non-black-earth, the industrial, and, partly, the northern black-earth gubernias.

The *Historico-Statistical Survey* (Vol. II) estimates that in the middle of the 60s, there were about 60 establishments with an output valued at about 270,000 rubles, while in 1880 there were 224 establishments with an output valued at 1,317,000 rubles. In 1890, according to the *Directory of Factories and Works* there were 192 establishments employing 3,418 workers, with an output valued at 1,760,000 rubles. "In the past 25 years," we read in the*Historico-Statistical Survey,* "the number of establishments engaged in starch production has increased 4 1/2 times and the total output 10 3/4 times; nevertheless, this productivity is far from covering the demand for starch" (p. 116), as evidenced by the increased starch imports from abroad. Analysing the data for each gubernia, the*Historico-Statistical Survey* reaches the conclusion that our production of potato starch (unlike that of wheat-starch) is of an agricultural character, being concentrated in the hands of peasants and landlords. "Showing promise of extensive development" in the future, "it is even now furnishing our rural population with considerable advantages" (126).

We shall see in a moment who enjoys these advantages. But first let us note that two processes must be distinguished in the development of starch production: on the one hand, the appearance of new small factories and the growth of peasant production, and on the other, the concentration of production in large steam-powered factories. For instance, in 1890 there were 77 steam-powered factories, with 52% of the total number of workers and 60% of the total output concentrated in them. Of these works only 11 were established before 1870, 17 in the 70s, 45 in the 80s, and 2 in 1890 (Mr. Orlov's *Directory).*

To acquaint ourselves with the economy of peasant starch production, let us turn to local investigations. In Moscow Gubernia, in 1880-81, 43 villages in 4 uyezds engaged in starch making. The number of establishments was estimated at 130, employing 780 workers and having an output valued at not less than 137,000 rubles. The industry spread mainly after the Reform; its technique gradually improved and larger establishments were formed requiring more fixed capital and showing a higher productivity of labour. Hand graters were replaced by improved ones, then horse power appeared, and finally the *drum* was introduced, considerably improving and cheapening production. Here are data we have compiled from a house-to-house census of "handicraftsmen," according to size of establishment: Thus we have here small capitalist establishments in which, as production expands, the employment of hired labour increases and the productivity of labour rises. These establishments bring the peasant bourgeoisie considerable profit, and also improve agricultural technique. But the situation of the workers in these workshops is very unsatisfactory, owing to the extremely insanitary working conditions and the long working day.

The peasants who own "grating" establishments farm under very favourable conditions. The planting of potatoes (on allotment, and chiefly on rented land) yields a considerably larger income than the planting of rye or oats. To enlarge their business the workshop owners rent a considerable amount of allotment land from the poor peasants. For example, in the village of Tsybino (Bronnitsy Uyezd), 18 owners of starch workshops (out of 105 peasant families in the village) rent allotments from peasants who have left in search of employment, and also from horseless peasants, thus adding to their own 61 allotments 133 more, which they have rented; concentrated in their hands are a total of 194 allotments, i.e., 44.5% of the total number of allotments in the village. "Exactly similar things," we read in the *Returns*, "are met with in other villages where the starch industry is more or less developed" (*loc. cit.*, 42). The owners of the starch workshops have twice as much livestock as the other peasants: they average 3.5 horses and 3.4 cows per household, as against 1.5 horses and 1.7 cows among the local peasants in general. Of the 68 workshop owners (covered by the house-to-house census) 10 own purchased land, 22 rent non-allotment land and 23 rent allotment land. In short, these are typical representatives of the peasant bourgeoisie.

Exactly analogous relations are to be found in the starch making industry in the Yuryev Uyezd, Vladimir Gubernia (V. Prugavin, *loc. cit.*, p. 104 and foll.). Here, too, the workshop owners carry on production mainly with the aid of wage-labour (out of 128 workers in 30 workshops, 86 are hired); and here, too, the workshop owners are far above the mass of the peasantry as far as stock-breeding and agriculture are concerned; they use potato pulp as feed for their cattle. Even real capitalist farmers emerge from among the peasants. Mr. Prugavin describes the farm of a peasant who owns a starch works (valued at about 1,500 rubles) employing 12 wage-workers. This peasant grows potatoes on his own farm, which he has enlarged by renting land. The crop rotation is seven-field and includes clover. For the farm work he employs from 7 to 8 workers, hired from spring to autumn ("from end to end"). The pulp is used as cattle feed, and the owner intends to use the waste water for his fields.

Mr. V. Prugavin assures us that this works enjoys "quite exceptional conditions." Of course, in any capitalist society the rural bourgeoisie will always constitute a very small minority of the rural population, and in this sense will, if you like, be an "exception." But this term will not eliminate the fact that in the starch-making area, as in all the other commercial farming areas in Russia, a class of rural entrepreneurs is being formed, who are organising capitalist agriculture.

### *Vegetable-Oil Production*

The extraction of oil from linseed, hemp, sunflower and other seeds is also frequently an agricultural industry. One can gauge the development of vegetable-oil production in the post-Reform period from the fact that in 1864 the vegetable-oil output had an estimated value of 1,619,000 rubles, in 1879 of 6,486,000 rubles, and in 1890 of 12,232,000 rubles. In this branch of production, too, a double process of development is to be observed: on the one hand, small peasant (and sometimes also landlord) oil presses producing oil for sale are established in the villages. On the other hand, large steam-driven works develop, which concentrate production and oust the small establishments. Here we are interested solely in the agricultural processing of oil-bearing plants. "The owners of the hempseed oil presses," we read in the *Historico-Statistical Survey* (Vol. II), "belong to the well-to-do members of the peasantry"; they attach particular value to vegetable-oil production because it enables them to obtain excellent feed for their cattle (oil cake).

Mr. Prugavin (*loc. cit.*), noting the "extensive development of the production of linseed oil" in the Yuryev Uyezd, Vladimir Gubernia, states that the peasants derive "no little advantage" from it (pp. 65-66), that crop and stock raising is conducted on a far higher level by peasants who own oil presses than by the bulk of the peasantry and that some of the oil millers also resort to the hire of rural workers (*loc. cit.*, tables, pp. 26-27 and 146-147). The Perm handicraft census for 1894-95 also showed that crop raising is conducted on a much higher level by handicraft oil millers than by the bulk of the peasants (larger areas under crops, far more animals, better harvests, etc.), and that this improvement in cultivation is accompanied by the hiring of rural workers. In the post-Reform period in Voronezh Gubernia, there has been a particular development of the commercial cultivation of sunflower seed, which is crushed for oil in local presses.

The area under sunflowers in Russia in the 70s was estimated at about 80,000 dess. (*Historico-Statistical Survey*, I), and in the 80s at about 136,000 dess., of which 2/3 belonged to peasants. "Since then, however, judging by certain data, the area under this plant has considerably increased, in some places by 100 per cent and even more" (*Productive Forces*, I, 37). "In the village of Alexeyevka alone" (Biryuch Uyezd, Voronezh Gubernia), we read in the *Historico-Statistical Survey*, Vol. II, "there are more than 40 oil presses, and Alexeyevka itself, solely owing to sunflowers, has prospered and grown from a wretched little hamlet into a rich town ship, with houses and shops roofed with sheet iron" (p. 41). How this wealth of the peasant bourgeoisie was reflected in the condition of the mass of the peasantry may be seen from the fact that in 1890, in the village of Alexeyevka, out of 2,273 families registered (13,386 persons of both sexes), 1,761 had no draught animals, 1,699 had no implements, 1,480 cultivated no land, and only 33 families did not engage in industries.

### *Tobacco Growing*

In conclusion, let us make some brief observations on the development of tobacco growing. The average crop in Russia for the years 1863-1867 was 1,923,000 poods from 32,161 dess.; for 1872-1878 it was 2,783,000 poods from 46,425 dess.; for the 80s, it was 4 million poods from 50,000 dess. The number of plantations in the same periods was estimated at 75,000, 95,000 and 650,000 respectively, which evidently indicates a very considerable increase in the number of small cultivators drawn into this type of commercial farming. Tobacco growing requires a considerable number of workers. Among the types of agricultural migration note is therefore made of migration to tobacco plantations (particularly to the outer gubernias in the South, where the cultivation of tobacco has recently expanded with exceptional rapidity). Reference has already been made in publications to the fact that the workers on the tobacco plantations lead a very hard life.

In the *Survey of Tobacco Growing in Russia* (Parts II and III, St. Petersburg, 1894, published by order of the Department of Agriculture), there are very detailed and interesting data on tobacco growing as a branch of commercial farming. Mr. V. S. Shcherbachov, describing tobacco growing in Malorossia, gives wonderfully precise information on three uyezds of Poltava Gubernia (Priluki, Lokhvitsa and Romny). This information, gathered by the author and arranged by the Bureau of Statistics, Poltava Gubernia Zemstvo Board, covers 25,089 peasant farms in the three uyezds that grow tobacco; they have 6,844 dessiatines under tobacco and 146,774 dessiatines under cereals. We see an enormous concentration of both the tobacco and the cereal area in the hands of the capitalist farms. Less than one-eighth of the farms (3,000 out of 25,000) hold more than half the area under cereals (74,000 dess. out of 147,000), with an average of nearly 25 dess. per farm. Almost half the area under tobacco (3,200 dess. out of 6,800) belongs to these farms, the average per farm being over 1 dessiatine, whereas for all the other groups the area under tobacco does not exceed one-to two-tenths of a dessiatine per household.

The owner of 7 dessiatines under tobacco must there fore have at least 14 workers; in other words, he must undoubtedly base his farm on wage-labour. Some grades of tobacco require not two but three seasonal workers per dessiatine, and day labourers in addition. In a word, we see quite clearly that the greater the degree to which agriculture becomes commercial, the more highly developed is its capitalist organisation.

The preponderance of small and tiny farms among the tobacco growers (11,997 farms out of 25,089 have *up to one-tenth* of a dessiatine planted) does not in the least refute the fact of the capitalist organisation of this branch of commercial agriculture; for this mass of tiny farms accounts for an insignificant share of the output (11,997, i.e., nearly half the farms, have in all 522 dess. out of 6,844, or less than one-tenth). Nor do "average" figures, to which people so often confine themselves, provide a picture of the real situation (the average per farm is a little over 114 dessiatine under tobacco).

In some uyezds the development of capitalist agriculture and the concentration of production are still more marked. In the Lokhvitsa Uyezd, for example, 229 farms out of 5,957 each have 20 dessiatines and more under cereals. Their owners have 22,799 dess. under cereals out of a total of 44,751, i.e., more than half. Each farmer has about 100 dess. under crops. Of the land under tobacco they have 1,126 dess. out of 2,003 dess. And if the farms are grouped according to area under tobacco, we have in this uyezd 132 farmers out of 5,957 with two and more dessiatines under tobacco. These 132 farmers have 1,441 dess. under tobacco out of 2,003, i.e., 72% and more than *ten* dessiatines under tobacco per farm. At the other extreme of the same Lokhvitsa Uyezd we have 4,360 farms (out of 5,957) having up to one-tenth of a dessiatine each under tobacco, and altogether 133 dessiatines out of 2,003, i.e., 6%. It goes without saying that the capitalist organisation of *production* is accompanied here by a very considerable development of *merchant's* capital and by all sorts of exploitation outside the sphere of production. The small tobacco growers have no drying sheds, are unable to give their tobacco time to ferment and to sell it (in 3 to 6 weeks) as a finished product. They sell the unfinished product at *half the price* to buyers-up, who very often plant tobacco themselves on rented land. The buyers-up "squeeze the small planters in every way" (p. 31 of cited publication). Commercial agriculture is commercial capitalist production: this relation can be clearly traced (if only one is able to select the proper methods) in this branch of agriculture too.

## THE JUDICIAL SYSTEM-SERVES THE INTERESTS OF THE RULING CLASSES

There are many people who believe that the institutionalisation of justice in the form of the judicial system that accompanied the emergence of bourgeois democracy only serves the interests of the ruling classes in this country. For them it is only natur al that the courts will not and cannot give justice to the poor. But for those of us who do believe that the courts-like any other institution in this troubled democracy-are contested arenas for conflicting interests, the only means of ensuring that the courts continue to function as institutions that affirm democracy, is by subjecting them to intense and persistent public scrutiny.

It is now a matter of widespread concern that the judicial system of this country (India) and its judgments are becoming increasingly anti-people: the Bhopal judgment as well as the one on the Narmada are both landmarks of justice denied. The recent judgment on the relocation of small industries and the fate of 25 lakh workers in Delhi becomes incomprehensible when you consider the government data that 67 per cent of all pollution in Delhi comes from vehicles. Yet there is no judgment on the sale, purchase or use of private cars, no real attempt to provide a better public transport system. The rich cannot be touched, but the industries and the workers must go. Does it not need to be asked then, what kind of justice is this, that is so divorced from its real ob jectives and what ends does it seek to meet?

The judgment on the Sardar Sarovar Project has to be scrutinised not only because of what it will mean for the millions of people in the Narmada valley who are being uprooted even while it has been made abundantly clear that there is simply no agricultural land available to rehabilitate them. It has to be scrutinised also because it draws larger conclusions about big dams in general (based not on empirical evidence, but on judicial conjecture) and about popular movements and people's access to the system of justice in particular

The majority judgment suggests that people cannot appeal to the courts of law once a project is under way. Since communities are never informed about a project until it begins to be executed, and since this judgment dec rees that they cannot appeal once it begins to be executed, does it mean that affected people can never question a patently bad project? This can only mean that the judicial system-one of the foundations of a democratic system-is unavailable to the common people of this country and their struggles (Front Line, Volume 17-Issue 26, Dec. 23, 2000-Jan. 05, 2001).

Living in a country where Neel Darpan, a celebrated play about the plight of the indigo planters during colonial times, inspired peasant struggles and a resistant subaltern consciousness; where the independence struggle (and many people's movements before and after) were marked by the participation of writers and artists.

Understand that the environmental battles of today and tomorrow are not just battles between the Indian elite and the peasantry or workin g class. The battle is between the large mass of common people in this country and global corporatisation. The fight over the control and the use of our lands and rivers is going to be as much in the forests of Madhya Pradesh (where tribal people oppose the World Bank Forestry Project), as on the streets of Andhra Pradesh (where farmers, energy workers and domestic consumers fight the wrecking of the power sector on IMF-World Bank prescriptions), as in the Narmada valley. You will have to decide which side you are on (C. Palit, 2001).

## TREE WORSHIP

Tree worship was a part of the religious faith in the prehistoric Indus civilization. The Vedas have praised trees as the sources of herbal medicine needed to fight diseases and for providing wood, as well as food in the pre-agricultural period.

In the Buddhist religion, tree worship had a special place as is evident from the discoveries in places like Sanchi, Barhut, Amaravati, Budhgaya where the mounds or stupas have revealed rich decorative relief work detailing scenes of tree-worship and images of "Brikha" or tree-god.

Trees being nature's major processors of solar energy which is vital for our existence, and yielding flowers, fruit, wood or medicine, have been worshipped by the ancient people of Indian Sub-Continent as a matter of gratitude. Manu believed that they were conscious like humans and felt pleasure and pain. Indian sages and seers eulogized asvattha or peepal (Ficus religiosa), gular (Ficus glomerata), neem (Azadirachta indica), bel (Aegle marmelos), bargad or banyan (Ficus bengalensis), asoka (Sereca indica), amala (Phyllanthus emblica), arjuna (Terminalia arjuna) and many other trees which acquired social and religious sanctity with the passage of time.

It is considered that primitive races of Bengal were tree-worshippers.. A number of trees have become objects of worship. In the popular belief, trees are seen to be favoured by different gods and goddesses who often lived in them.

It was this belief that Marmelos tree to Shiva, Tulsi (Basil) to Bishunu, Shal to Durga and Shij to Manesa. In Bikrampur in the District of Dhaka, Hindu women have been worshipping a particular Neem tree by daubing it with vermillion and oil as they think the godess Kali lives in this tree.

Bono bibi or Lady of the Forests-is the presiding female detity of the Sunderbans (Mangrove Forest) cultural zone.

She is the gurdian deity of the forest. Both the Hindu and Muslim communities pay their respects before venturing into the forest. Also known as Basuli, Bibima, Bon Durga or Bon Kali, she is potrayed by clay molders either mounted on a tiger or a hen. She is pretty and graceful, ever eager to protect the people of the Snderbans.

Our ancestors worshipped the elements: the sun, earth, water, wind, thunder and lightning. The ritual abides; the spirit is gone. We still regard the peepal sacred because the Buddha gained enlightenment meditating under its branches-hence the Latin name ficus religiosa. Its cousin banyan or barh is still worshipped in villages across the country. So is the tulsi (Basil) grown and worshipped in millions of Hindu homes. We worship trees but we do not look after them. We cut down forests every day to cremate our dead. We use wood as fuel to cook and keep ourselves warm. We deprive birds and animals of food and shelter. We must reverse the process, learn to love and cherish our trees.

Berholt Brecht captures man's yearning to establish a close relationship with a tree:

*Morning Address to a Tree Named Green*

*Green, I owe you an apology*

*I couldn't sleep last night because of the noise of the storm.*

*When looked out I noticed you swang*

*Like drunken ape. I remarked on it.*

*Today the yellow sun is shining in your bare branches*
*You are shaking off a few tears still, Green.*
*But now you know your own worth.*
*You have fought your bitterest fight of your life.*
*Vultures were taking an interest in you.*

*And now I know:it's only by your inexorable*
*Flexibility that you are still upright this morning.*
*In view of your success it's my opinion today:*
*It was no mean feat to grow up so tall*
*In between the tenements, so tall, Green, that*
*The storm can get at you as it did last night.*
*Worship of Tree 'Karam Puja'*

With a view to starting the festival the 'Thakur' (priest) along with some of the members of their community went to Jonepur, some two kilometres away from Natshal, to cut a branch of Karam or Kadamba tree. There they lighted an earthen lamp (Pradip) and offered worship at the foot of the tree. Then one of them climbed the tree and cut a branch of it. They returned to Natshal, one of the venues of the festival with that branch of the Karam tree and planted it.

The aborigine men and women passed the whole night by singing and dancing surrounding the branch of the 'Karam' with 'Madal' and 'Karatal'. In the morning, they sank the branch in a nearby pond. This was the main ritual the aborigines had long been performing. But there is a story that they believe to be the cause of introduction of Karam Puja.

The aborigines, who live mainly on agriculture, believe that to get proper benefit from agriculture they must worship the branch of Karma (Kadamba) in the name of the 'Karma God'.

The story that they believe is like this: Karma and Dharam were two brothers. Karma worked hard but Dharma did not work. He only worshipped a branch of a tree.

At this, being very angry Karma once threw away the branch which fell on an island across seven seas and thirteen rivers. Karma began to suffer for his neglect of the Kadamba branch and found no more success in agriculture. Karma realised his guilt and after toiling too much took back the branch and started worshipping it. At this he regained his success in agriculture. From that moment 'Dal Puja' or 'Karam festival' came in culture of the aborigines.

Karam festival was actually the festival of the 'Orao' tribe who used to celebrate the festival at their respective areas. Jatiya Adibasi Parishad and Adibashi Sangskritik Parishad jointly started celebrating the festival about eight years ago at Natshal field on the next day of the main ritual. Now it has become a great communion of all the aborigines like Orao, Santal, Munda, Mahato and Raichatri.

Sounds of 'dhol' and clapping of the aborigine men, women, old, young and children create a dancing excitement in the blood of all gathered there. But it was closely observed that a section of political personalities have spread their claws to take the minority group under their control. They make the total arrangement of the festival at Natshal field from background although they do not belong to that community (The Independent, September 4, 2004).

Yusuf et. al., (1994) in a recent publication gave a list of 546 medicinal plants that occur in Bangladesh. However, the inventory is not complete, and many plants with medicinal value are yet to be discovered.

The Rangamati Hill District in CHT still harbors a portion of virgin forest. But the procurement of medicinal plants from the wild habitat for professional collectors to make local medicines is unscientific, indiscriminate, and in most cases leads to overexploitation. There is severe depletion of the natural stands, without any provision for the regeneration of species. Some rare species like Ulat chandol (Gloriosa superba), Sarpo gandha (Rawolfia serpentina), and Aswa gandha (Withania somnifera) have become regionally endangered.

Prior to the last two centuries medical practitioners-whether allopaths, homeopaths, naturopaths, herbalists, or shamans-have to know the plants in their areas and how to use them, since many of the drugs were derived from the plants. Plants contain compounds that include a pharmacological reaction in the human body. Plants are very rich in secondary compounds including alkaloids, glycosides, essential oils, and other organic constituents, are responsible for the medical qualities of plants. Alkaloids obtained from vascular plants are among the most important pharmacologically active compounds. They are bitter-tasting organic compounds that are basic (alkaline) in their chemical properties. In Bangladesh and India many bitter spinach varieties that grow in water and land are known as blood cleaning agent and are preferred as appetiser for thr hundreds of years or more.

Man is not only a great inventor and builder, but he has also proved to be the most destructive force ever to appear on the face of the earth. Besides less than ten percent of the population of this planet enjoys all the resources and determines the future course. Statistics show that the devastation of vast areas of remaining undeveloped surface of our world have been destroyed with inevitable extinction of thousands of plant and animal species.

People have recognised the medical value of plants for thousands of years. In the Vedas, which stretch back more than five thousand years mentioned that spices are not only an integral part of culture but also invaluable to cure for every ailment known to man. Even though our earliest ancestors may not have understood how or why certain plants cured specific ailments, they were well aware that plants heal as well nourish. In developing countries today the majority of the population rely on herbal drugs. About two thousands plants are used medically in the Indian-Subcontinent, while three-quarters of the population of China still use herbal medicine. Before synthetic chemicals dominated medicine, as they do to day, roughly 80 per cent of all drugs were derived from plant materials. Chemists eventually developed synthetic version of many drugs, but these man made products would never existed without nature leading the way. In some cases, chemists have not yet learned to duplicate nature. There are stillmany unknown wild tropical plants of our botanical heritage not yet researched or discovered and many potential cures.

## JHUM CULTIVATION

The people of North East India and hilly areas of Bangladesh represent a fascinating variety of cultures. Jhum plays an important cultural role in local customs, traditions, and practices, besides offering economic security to farmers. It would be unfortunate if developmental programmes based on misguided opinions about jhum suppress this unique form of agriculture. Only occupations providing monetary and social benefits perceived by jhumias to outweigh the cultural and security benefits embodied by jhum are likely to gain acceptance. A balanced approach to development that also recognises the merits of jhum is needed. Then, this remarkable form of organic farming may persist into the 21st century.

Jhum as commonly practised by indigenous tribes in North East India. This 'primitive' form of agriculture, according to supporters of "deforestation":

- resulted in serious environmental problems:
- loss of forest cover, erosion of topsoil, desertification, and
- declines in forest productivity.

Others have also decried jhum as an inefficient form of agriculture, an impediment to progress of forestry, and an agent of destruction of biodiversity. Such beliefs have been widespread since British times, and have even resulted in forcible suppression of the practice, oppression and relocation of tribals in

Central India and other hill regions. Rapid demographic and social changes have occurred in many tribal societies of North East India. The environmental impacts of jhum cultivation and its role in people's lives have concurrently changed. The conversion of over 80% of the population to Christianity in less than a century (1894-1994) has dislodged the significant role of superstition and mystique in peoples' relationship with their natural environment. A large majority of peoples is tribal and dependent on jhum for its subsistence and livelihood.

## Advantage of Jhum Cultivation

- In contrast, studies by ethnologists have tended to view shifting cultivation favourably. It is considered a diversified system, well adapted to local conditions in moist forest and hilly tracts.Others have argued that traditional shifting cultivation may not be as destructive as modern forest exploitation for timber. Clearance of small patches of forest with long fallow periods may even enhance biodiversity in the landscape due to the creation of a variety of habitats. Amidst such contrasting views, there is a clear need for reliable empirical and scientific data on the nature and ecological impact of jhum.
- Jhum cultivation usually involves cutting of second-growth bamboo forests. Since old growth or primary forest is less extensively available and is more difficult to clear, they are cultivated infrequently. The clearing work usually begins in January-February. The slashed vegetation is allowed to dry on the hill slopes for 1-2 months prior to burning in March-April. Crops are sown with the first rains in April in plots that are 1-4 ha in area. Usually, inter-cropping of one or more paddy varieties with 15-20 other crops (vegetables, maize, chillies, gourds, cotton, arum, and mustard) is carried out.
- Studies showed that, far from being primitive and inefficient, jhum is an ingenious system of organic multiple cropping well suited to the heavy rainfall areas of the hill tracts. The economic and energetic efficiency of jhum is higher than alternative forms of agriculture such as terrace and valley cultivation. This is mainly because terrace and valley cultivation needs expensive external input such as fertilisers (which often get leached or lost in the heavy rainfall hill slopes) and pesticides, besides labour for terracing.
- The superiority of jhum cultivation over some forms of sedentary cultivation partly explains the persistence of this form of agriculture in North East India. Other reasons include the economic security provided by jhum and its cultural importance to indigenous tribes. Poor access

to markets, capital, and technical knowhow of more commercially rewarding alternatives such as horticulture and cash crop cultivation also hinders the transition to other occupations. Clearly, one cannot do away with jhum assuming it to be a primitive and inefficient system, as attempted in governmental jhum control programmes and new land use policies. Instead, an unbiased understanding of the advantages of jhum is required for proper design and implementation of developmental programmes.

- Erosion of valuable topsoil in the hills due to jhum has been alleged to cause siltation and floods in the plains. Singh has reviewed studies carried out by the Indian Council of Agricultural Research that compared soil erosion from jhum fields with other forms of cultivation on terraces and contour bunds. These studies show that jhum fields cultivated for a single year and abandoned (the most common practice) have less erosive losses of soil than the other forms of settled cultivation.
- Soil erosion is minimised in jhum due to the retaining of rootstocks of bamboo and trees in burned plots, the rapid recovery of weeds and bamboo following abandonment, and the interspersion of forests and fields on hill slopes. The evidence for siltation of rivers and floods because of soil erosion due to jhum is weak and possibly untenable. Other factors, such as large scale logging for timber extraction, may be responsible to a greater extent for the deforestation and environmental problems in North East India.

## CORIANDER CULTIVATION

### Poor People's Rich Food

Indian turnip has a round, flattened, perennial, rhizome (cormus), the upper part of which is tunicated like the onion, the lower and larger portion tuberous and fleshy, giving off numerous long, white radicles in a circle, from its upper edge; the under side is covered with a dark, loose, wrinkled epidermis The spathe is ovate, acuminate, convoluted into a tube at the bottom, flattened and bent over at the top like a hood, varying in color internally, being green, dark-purple, black, or variegated with pale-greenish stripes on a dark ground, supported by an erect, round, green, purple, or variegated scape, invested at the base by the petioles and their acute sheaths.

The plant has one enormous leaf and one spadix annually. It requires hand pollination in Britain. When ripe for pollination, the flowers have a foetid smell to attract carrion flies and midges. This smell disappears once the flower has been pollinated.

The Arum family, Aroidae, which numbers nearly 1,000 members, mostly tropical, and many of them marsh or water plants, is represented in this country by a sole species, Arum maculatum (Linn.), familiarly known as Lords and Ladies, or Cuckoo-pint.

**Description:** The flowering organs are contained in a sheath-like leaf called a spathe, within which rises a long, fleshy stem, or column called the spadix, bearing closely arranged groups of stalkless, primitive flowers.

The Arum has large tuberous roots, somewhat resembling those of the Potato, oblong in shape, about the size of a pigeon's egg, brownish externally, white within and when fresh, fleshy yielding a milky juice, almost insipid to the taste at first, but soon producing a burning and pricking sensation.The acridity is lost during the process of drying and by application of heat, when the substance of the tuber is left as starch. When baked, the tubers are edible, and from the amount of starch, nutritious. This starch of the root, after repeated washing, makes a kind of arrowroot, formerly much prepared in the Isle of Portland, and sold as an article of food under the name of Portland Sago, or Portland Arrowroot, but now obsolete. For this purpose, it was either roasted or boiled, and then dried and pounded in a mortar, the skin being previously peeled.

This starch, however, in spite of Gerard's remarks, forms the Cyprus Powder of the Parisians, who used it as a cosmetic for the skin, and Dr. Withering says of this cosmetic formed from the tuber starch, that 'it is undoubtedly a good and innocent cosmetic'; and Hogg (Vegetable Kingdom, 1858) reported its use in Italy to remove freckles from the face and hands.

In parts of France, a custom existed of turning to account the mucilaginous juice of the plant as a substitute for soap, the stalks of the plant when in flower being cut and soaked for three weeks in water, which was daily poured off carefully and the residue collected at the bottom of the pan, then dried and used for laundry work.

**Constituents:** The fresh tuber contains a volatile, acrid principle and starch, albumen, gum, sugar, extractive, lignin and salts of potassium and calcium. Saponin has been separated, also a brownish, oily liquid alkaloid, resembling coniine in its properties, but less active. Arum leaves give off prussic acid when injured, being a product of certain glucosides contained, called cyanophoric glucosides

The dried root was recommended as a diuretic and stimulant, but is no longer employed. The British Domestic Herbal describes a case of alarming dropsy with great constitutional exhaustion treated most successfully with a medicine composed of Arum and Angelica, which cured in about three weeks.

A homoeopathic tincture is prepared from the plant, and its root, which proves curative in diluted doses for a chronic sore throat with swollen mucous membranes and hoarseness, and likewise for a feverish sore throat.

An ointment made by stewing the fresh sliced tuber with lard is stated to be an efficient cure for ringworm, though the fresh sliced tuber applied to the skin produces a blister. The juice of the fresh plant when incorporated with lard has also been applied locally in the treatment of ringworm.

Arum, an esculent edible root, though, trifled literally, used to be considered an occasional vegetable in a Bengali household's food menu. Arum is the only vegetable that survived the flood and rain although farmers in some places lost their produces to the deluge and the downpour. This was the season of cabbage and cauliflower but those had been either washed away or damaged (India-Bangladesh Flood 2004). Arum can survive under water longer than other agricultural products, so it has become the only hope for consumers (T. Maji, October 18, 2004).

The genus Arum (Araceae) is represented by some 20 taxa in Turkey. Having tuberous roots, broadly hastate vigorous leaves, greenish-yellow spathes A. italicum grows in northern Turkey and flowers between April and May and its reddish berry type fruits ripen in July. Containing significant amount of calcium oxalate crystals, oxalic acid and oxalates in addition to volatile and/or easily destroyed irritating substances, Arum taxa are toxic. However, dried or fresh parts thereof are used for food and in folk medicine in Turkey. Tubers and ripe fruits are used in the treatment of rheumatism and hemorroids while the leaves are consumed as a food.

Arum calocasia (Arbi)It is cool, give strength, an appetizer and increases the quantity of milk in mother's breasts It is diuretic causes the formation of excessive wind and phlegm in the body. It increases the quantity of semen, cures plethora and dysentery.

**Uses:**

- It is mostly used as a vegetable. Although there are many types of arum calocasia, their properties are approximately the same.
- Grinded tender leaves of arum calocasia mixed with powdered cumin seeds cures excessive bile in the body.
- Vegetable of arum calocasia increases the quantity of mother's milk.

Arisæma curvatum, Hook.; Kunth. India: roots eaten.

Although no specific mention has been seen for this species, it belongs to a family where most of the members contain calcium oxalate crystals. This substance is toxic fresh and, if eaten, makes the mouth, tongue and throat feel as if hundreds of small needles are digging in to them. However, calcium oxalate is easily broken down either by thoroughly cooking the plant or by fully drying it and, in either of these states, it is safe to eat the plant. People with a tendency to rheumatism, arthritis, gout, kidney stones and hyperacidity should take especial caution if including this plant in their diet

The root is carminative, restorative, stomachic and tonic. It is dried and used in the treatment of piles and dysentery. The fresh root acts as an acrid stimulant and expectorant, it is much used in India in the treatment of acute rheumatism.

Millions of people in eight districts in Bangladesh are on the brink of starvation. Thousands more face the threat of ill-health and unemployment. Although the government says it has initiated relief measures, the ground realities belie this claim.

One of South Asia's most severe droughts, coupled with a 400% increase in the price of essential goods, has left over two million people in north-western Bangladesh on the brink of starvation and forced residents in eight districts to migrate in search of food and employment.

Famine stricken people looking for the Arum roots but this has become rare and expensive (October, 2004).

A near-famine situation the northern districts-Story of Rafique and Others

Arum dracunculus, L. France: starch of root recommended as a famine food for extending bread flour, after removal of acrid element. Ref. PARMENTIER.

Calla palustris, L. France: starch of root recommended as a famine food for extending bread flour, after removal of acrid element. Sweden: unidentified part of plant used in preparation of bread. Bengladesh: greens eaten; roots may be boiled with rice or cooked as curry and may contain chemical component which can irritate mouth and throat. Vernacular name: Water Dragon. Bangladesh: Kachu. Ref. DARLINGTON & AMMAL, DILLINGHAM (1900), PARMENTIER, RAHAMAN.

Colocasia esculenta Schott (syn.Caladium esculentum, Vent.; Colocasia antiquorum, Schott.) (Shortt gives "Callादem esculuntum," the genus and species probably being misspelled). India (Madras Presidency): leaves and leaf-stalks eaten as greens. Kapingamarangi: leaves of the wild taro eaten. Vernacular names-Tamil: Sainmay keeray, Shamay kilangu. Telugu: Chama kura, Chama dumpa;

Pistia stratiotes, L. India: used as a famine food in 1877-1878. Herb is recorded as eaten at other tjmes. China: young leaves eaten cooked. Philippines:

used to treat gonorrhea. Plant has a high potash content, and contains stinging crystals. Occurs in great abundance on the surface of stagnant water and slowly-moving streams;

Chemical Composition.:In addition to its acrid principle it contains a large proportion of starch; also, gum, albumen, saccharine matter, calcium and potassium salts, and extractive. When the acrid property is driven off by heat, the root yields a pure, delicate, amylaceous matter, resembling the finest arrowroot, very white and nutritive. That raphides of oxalate of calcium give to the corm its acridity has been asserted by Weber (1991).

Medical Uses:Recommended in flatulence, croup, whooping-cough, stomatitis, asthma, chronic laryngitis, bronchitis, pains in the chest, colic, low stage of typhus, and various affections connected with a cachectic state of the system. Externally it has been used in scrofulous tumors, tinea capitis, and other cutaneous diseases. Its action in the prostration of low fevers with wild delirium is due to its effects upon the cerebral centers. It is reputed useful in cerebro-spinal fever and scarlatina, when delirium is present, when the tongue is swollen, red, and painful, and the buccal membranes inflamed.

Chronic laryngitis, or minister's sore throat, with sudden hoarseness and aphonia, is specifically influenced by arum. It is also useful in ulceration of the larynx and pharynx. It is a good remedy, internally and locally, for aggravated red sore throat. The powdered root may be given in 10-grain doses, increased, if required, to 20 or 30 grains, and repeated every 3 or 4 hours. It may be taken in sweetened mucilage, syrup, or honey. Specific arum, 1/10 to 10 drops. Its specific effects are best obtained by minute doses of the specific arum-1/10 to 1/2 drop doses.

## Arum Farming gains Ground in Bangladesh

Arum farming has been gaining popularity in all the seven Upazillas of the district, as this cultivation is bringing profits to cultivators. Previously arum was found in markets of the district and some markets of the other districts. Only a little quantity of arum was supplied to the markets. But now arum is available in the market. Now in all the haats and Bazaars arum is available, as the farmers have started its farming on commercial basis. Talking to Karim Mondol, farmer of village Par of Kendua Union under Sadar Upazilla said he produced, at least, 50 mounds of arum on his five decimal lands. Of the total, he has already sold 15 mounds in the markets at the rate of Taka 280 to Tk 300 per mound. According to Mondol, he will be able to earn Taka 7000 from his products. He had spent about Taka 1000 for purchasing seeds, preparing lands and for other reasons.

Farmers said they usually cultivate seven varieties of arum, which include 'man kachu, pani kachu, gut kachu, kalika kachu, bish kachu, ole kachu and panchamukhi kachu (locally known mukhi)'. Farmers sow arum seeds in the Bengla months of Jaistha and Asar. Cultivation of arum is not very difficult, as the crop needs neither fertilizers nor pesticides. Moreover, the soil and climate of the district is suitable for arum farming (July, 2004).

## Home Gardens-Stabilty of Ecosystem

Home or kitchen garden system proposed by Gonzalez (1985) and Allison (1983) is one of the agro-ecosystem that seems to be well adapted ecologically to tropics. Such gardens existed in this sub-continent (India) but due to reduction of land by farmers year to year and introduction of industrial/ pharmaceutical products from the cities this valuable heritage is now gradually disappearing.

Tropical homegardens with their large crop species and varietal diversity are regarded as an ideal production system for in situ conservation of plant genetic resources. They are also known to be fields of experimentation and domestication of wild plants. However, garden diversity varies according to ecological and socio-economic factors and/or characteristics of gardens or gardeners

A home garden with an overstory of trees and an understory of a mixture of herbs and small trees permits year-round harvesting of food products, as well as wide range of other products used by the local people, such as firewood, medical plants, spices and ornamentals. Relatively high species diversity provides resource-conserving and ecological sound farming system.

## Homegardens as a Special Agroforestry Niche for Women

The cultivation and management of homegardens by women is a widespread phenomenon among settled groups the world over (Buch 1980, Niñez 1985). This is particularly pronounced in Latin America in areas where women do not traditionally till the land, since it provides an agricultural production niche that is seen as an extension of the home. The homegarden is often a way around taboos against tilling the main cropland, and is usually considered an extension of the home as the women's domain. Moreover, by definition such plots are location-specific to the home area, and as such are accessible to women whose mobility may be limited by custom, or by the complex logistics of mixing travel with child care, food processing and food preparation. Homegardens provide an opportunity to intensify labor inputs to increase production, without adding time away from home and within a flexible schedule shaped around other household responsibilities (Chaney and Lewis 1980).

Summary of advantages of commons plantings for women:

- access to land for production
- access to better quality land than they would normally have access to the convenience (efficiency) of local access to concentrated plantings of normally scattered resources "one stop shopping"
- reduction and improved timing of labour inputs, e.g. through rotational labour
- economies of scale through easier fencing, maintenance, protection and marketing of products from concentrated block plantings
- concentrated access to training and assistance
- benefits of the "group learning curve"
- access to credit.

## Home Garden benefits Poor People

The people of Domar Upazila; Nilphamari, Bangladesh benefited from afforestation program, besides maintaining ecological balance has created job opportunities for poor people in the upazila. The members of the Samities (committe) look after the saplings and nurture them till they get matured.

The Samity members are given 40 per cent of the money earned from the sale. After cutting down the Bogra trees, saplings of ten percent medicinal tree and 25 percent fruit bearing tree are again planted. Again, the saplings of flower bearing trees are planted. The process goes on by rotation. At present, trees of different species on both sides of the railway line are increasing the natural beauty. Soil erosion had completely been stopped and the programme had created job opportunities for the rural people. Besides, the government is earning huge revenue (The Independent, November 13, 2004).

Home gardens are considered to be:

- Variable in size and design;
- Respond to local soil type, drainage patterns, cultural preferences, economic standing of the family, family size and age pattern reflecting a multiplicity of both ecological and cultural components;
- Flexible, dynamic, and changing, depending on the needs of the family.

Distribution of plants:

- Low diversity, regularly patterned planting of crops of potential cash values;

- High diversity, irregularly patterned planting of trees, shrubs, herbs etc;
- Low diversity, widely spaced planting of trees, with low grass or bare soil,
- Very high diversity, intercropped planting of ornamental herbs and shrubs,
- Moderate diversity, alternately planted fencerow surrounding the property primarily composed of fruits and fire wood tree species.

Homegardens are more reliable than crops fields for growing trees and vegetables and are important sources of income for the farmers of Bangladesh.

## Homegardens are more Reliable than Crops Fields for Growing Trees

Home garden represents the blending of knowledge gained by ecologists studying the dynamics and stability of tropical ecosystems with the knowledge of farmers and agronomists on how to manage the complexities of food producing ecosystems.

Two parallel systems of production forestry exist in Bangladesh: government forests managed by the Forest Department (FD) and privately owned homegardens. Of the country's total land area, about 1.48 million hectares (ha) are designated as government forest land that covers both natural and plantation forests. About 0.72 million ha of land are disignated as unclassified state forests under the control of the Ministry of Land. Homegardens constitute 0.27 ma he and are scattered all over the country. The public forest land, un-classed state forests and homegardens together make up about 17% (2.46 million hectares) of the potential tree growing area of the country the lowest figure of any South Asian country.

From the physical and socio economic points of view, homegardens are more reliable than crops fields for growing trees and vegetables and are important sources of income for the farmers of Bangladesh. It is observed that farmers tend to sell cropland to fight against pauperization, but retain their homegardens unless absolutely unavoidable: Even functionally landless farmers have their own homegardens, where they grow the essential commodies for subsistanc. It is observed that over half of the fruits, vegetables and spices grown in the homegardens are sold to meet family expenses. In Bangladesh farmers spent only 4.8-12.2% of their total labour. In homegarden management, but 26% to 47% of the total family expenses are met from selling homegarden products. During the last 40 years. the relative importance has shifted from the traditional forestry (in the government managed forests) to homegardens in such a way that today about 55% of requirement of timber, fuelwood and bamboo are met from the homegarden sources.

Sunderbans, the largest mangrove forest of the world contain many traditional medical plants that can be planted in many wetland areas of Bangladesh:

## Species with Possible Development Potential for Homegarden Use

1. Extinction of Crocodiles
2. Migratory and other Birds in Bangladesh in Danger
3. The Wood Sandpiper Tringa glareola Threating to Extinct

A terrain full of verdant trees, plants, herbs and foliage, the Sunderbans is one of the largest intact mangrove forests in the world declared a World Heritage Site by UNESCO in 1997. Thus globally, the Sundarbans is one of great importance. Home to a variety of species, the forest is unique in that many of its plants and animals are not found anywhere else in the world. Over many years land grabbers have harmed this rare ecosystem; and in this way transformation of the Sundarbans from jungles of great biodiversity to wet rice paddy fields occurred causing much damage to its resources.

Over 60 per cent of Sundari trees are dying in the Sundarbans mangrove forest with high salinity prevalent in Khulna and Jessore regions due to severe lack of sweet water flow from upstream points coupled with negative impact of the Farakka Barrage, a leading water expert said. Prof. Ainun Nishat, Country Director of IUCN, told BSS today that water is being withdrawn in the upstream of Farakka Barrage in the Uttar Pradesh, northern region and Bihar. He said lack of sufficient water not only hampers cultivation but also creates negative impact on fish resources in the rivers. Prof. Nishat said Sundari wood is more valuable than normal wood. Lack of sweet water contents in the Sundarbans mangrove forest kills Sundari trees, he said. He suggested that the country should ensure augmented flow of the Ganges river and this should be diverted to Khulna region.

Hydrologists told BSS that the water level fell by three feet at Hardinge Bridge point in December 2004. They said similar fall in the water level was also noticed at Gorai railway bridge point. The hydrologists of Bangladesh Water Development Board (BWDB) said Bangladesh received 145,000 cusec water at the Hardinge Bridge point on December 1, 2004.

They referred to the field reports and said the country received 107,000 cusec water on December 18 which means that the volume of fall stands at 38,000 cusec further. The BSS roving correspondent observed that the water levels in the Padma and the Gorai are falling continuously. Many shoals (chars) have emerged on the waterways of the Padma between Paturia and Daulatdia ghat. Continuous dredging is underway to ensure navigability of the waterway. The fall of water level has caused disruption in the ferry service (The Bangladesh Observer, January 04, 2005).

## Tannin

Ceriops decandra, Bruguiera gymnorhiza, Bruguiera parviflora, Rhizophora musconata, Xylocarpus decandra are the valuable trees that produce tannin in the Sunderbans. Studies by the Forest Research Institute, Dhera Dun, India have shown that the spray-dried extract of a blend of ceriops, myrobalans and Acacia nilotica bark contain 65 percent tannin and the blend is suitable for the manufacturing of crust leather. The use of mangrove bark or extract in tanning is locally well-known. and is not commercially used.

The small leather industry of Indian-subcontinent developed Indian vegetable tanned crust over a hundred years ago to preserve the hide in the safest way to suit Indian conditions. The development of leather processing industry was started in Bangladesh in the late 1940s. Until mid 1960s, the leather was dominated by vegetable tanned products for supply to W. Pakistan, Iran and Turkey.

Manufacture of wet blue, the chrome tanned semi-processed leather started featuring in 1965. There was a rapid growth of tanning industry in Bangladesh during 1970s and by the end of 70s. In 1999 Bangladesh exported leather and leather goods worth US $ 225 million. Now chrome-tanned processed leather is the shooting star of the export industry at the cost of serious environmental depletion with cancer producing substances.

There is a potential market for vegetable tanned leather products in the industrial countries. Development countries should not destroy their environment for export industry.

Since millions of years from the Himalayas to the dynamic coastal plain of Bengal was rich in panoramic vegetation and wild life. These tropical moist forests were botanically amongst the richest in the Indian sub-continent. The forests are most important as a repository of one of the world's richest of biodiversity

Sunderbans, the largest mangrove forests of the world, was once covered all along the coastal plain of Bangladesh. Had it been maintained, the Bay of Bengal would have turned into one of the largest fish grounds of the world, gained land one third to the present size of Bangladesh, and have protected millions of lives during cyclone storms. The problems of deforestation is mainly political and it can be solved, if poverty focused projects contain the attitude of "by the people and for the people" participation:

1. Our Blue Planet: Extinction of Mangrove Forests
2. Polluted Leather Industry and slums of Bangladesh
3. Poultry feed churned out from tannery waste
4. Plunder of forest resources unabated in Rangamati
5. HILSA Tenualosa ilisha King of Fishes-Going to Extinct?

## Neem Azadirachta indica–The Wonder Plant

Sitala Puja-Caitra navaratras: goddess Sitala who is said to reside in the neem tree is propitiated ritually; Pat Gosain festival in Bengal means neem tree worship; neem leaves are eaten on Vaisakha sukla saptami.

"To the best of my knowledge, no plant material with greater activity against abroader spectrum of pest insect species, has yet been found." Dr Martin Jacobson of United States Dept. of Agriculture-Agricultural Research Center in Beltsiville, Madison, USA.

The Neem is being heralded as a tree for solving global problems by the U.S. Department of agricultural. Equivalent products to NeemHit are already registered in U.S.A. (since 1992) and numbers of European countries.

The tree has relieved so many different pains, fevers, infections, and other complaints that it has been called "the village pharmacy."

"Azad dhirakat " from the Persian means "Excellent Tree, Noble Tree" referring to the usefulness and the considerable economic importance of the genus. Locally named in Bangladesh as nim, in In dia as nimba, nimuri etc., Nepal as nim, Tibetan as nimpa, traditionally used to make medicine and pesticid1es. Prof. Heinrich Schmutterer, Department of Phytopathology and Entomology working since thirty years on Neem tree and termed, "Neem is the one of the most fascinated trees of the world". In Bangladesh villagers brush their teeth with the Neem branch.

The neem tree (Azadirachta indica) is a tropical evergreen related to mahogany. Native to east India and Burma, it grows in much of southeast Asia and west Africa. A few trees have recently been planted in the Caribbean and several Central American countries

The neem tree has such a variety of medical applications that it is sometimes referred to as the village pharmacy. Now modern research is proving what has been long known by Ayurvedic medicine practitioners: neem is one of the most effective plant medicines in the world. An extremely powerful blood purifying agent and detoxicant, neem is also effective in the treatment of fever, malaria, skin diseases, dental problems, diabetes, tumors, arthritis, and jaundice. It has gained particular attention from scientists seeking a cure for AIDS, not only for its antiviral properties, but also because it boosts the immune system on all levels without destroying beneficial bacteria, unlike synthetic antibiotics.

Azadirachtin being the key molecule, more concentration on Research & developments have been targeted on Azadirachtin only in India and abroad. However now it clearly known that besides Azadirachtin, salannin, gedunin, azadirone, nimbin, nimbidine, nimbicidine, nimbinol, etc.. are also important liminoids which play an excellent synergistic effects on Insects/Pests.

Tree is considered a good purifier of air, due to its large leaf area. Native of Burma but grown all over Indian Sub-Continnet. Oilcake, obtained from seeds, is used as a fertilizer and manure. Green twigs are used as tooth brushes for cleaning teeth, and as a prophylactic for mouth and teeth complaints. Parts of the plant are used medicinally and the leaves are placed in suit cases to repel insects and to preserve woollens. An extract of leaves is used in tooth pastes and soaps Seeds yield famous margosa oil of disagreeable garlic like flavour. Oil is said to be effective in treatment of leprosy and skin diseases. Also used as a cure for manage in dogs. Leaves in poultice are used for healing of wounds. Ripe fruits are edible. Due to its bitter taste and disagreeable odour, not removed by conventional methods, neem oil has not been utilised on an industrial scale

Oil obtained from neem seed has been found to be suitable for soap making and for hydrogenation. Seed oil is also used as antiseptic and for burning purposes. Stones from fruits are used as beads in rosaries and necklaces. Azadirachtin, a substance isolated from the tree, has been found to have insect repellent and insecticidal properties. Bark yields tannin. Gum exudate from the bark is used in medicines as a stimulant, and for dyeing silk. Bark is useful in fever, nausea, vomitting and skin disease. Bitter principles of neem oil are reported to have been obtained by extraction with alcohol

The main component of the oil is nimbidin which is very bitter. Nimbidin is used for making several pharmaceutical preparations including emulsions, liquors, ointments, medicinal cosmetics such as lotions, shampoos, creams, hair tonics and gargles. Timber is used for agricultural implements and furniture.

In addition to its numerous uses as a healing agent, neem has been receiving much attention for the ecological benefits it provides. For centuries Indians have been mixing neem leaves with stored grains to prevent insect infestation. But neem is not simply a natural alternative to pesticides;

increasingly it is being used to reverse desertification and to reduce erosion and deforestation, making it an important weapon in the fight against global warming. Neem's many practical applications make it of enormous interest to anyone concerned about health and ecology. Included are recipes and practical tips that let you enjoy the many benefits of this miraculous plant.

## OTHER PRODUCTS

Neemaura is a environmentally friendly natural Neem bitterns,non toxic and safe biodegradable urea coating agent containing Neem Triterpenes inhibit the growth of nitrifying bacteria resulting in delayed transformation of ammoniacal nitrogen into nitrite nitrogen. This ensures slow and continuous availability of nitrogen matching the requirements of crop plants during their life cycle and effectively retards the nitrification of urea. Neemaura coated urea mineralizes much slower than plain urea at least two to three times under soil conditions by controlling the multiplication process of soil borne bacteria like, Nitrosomanas and Nitrobacter which are responsible for nitrification. Neemaura formulation contains neem bitters and sulphurous compounds, which are mainly responsible for retarding the process of bacterial action and protects urea from leaching, volatilization and also protects crop from insect pest result in higher yields.

### Biogrow

Bio-Organic soil enricher made from complete biodegradble organic Neem cake which is manure for green earth. Neem cake is rich in sulphur compounds, in addition to its intrinsic N.P.K. value, it possess bitter terpenoids such as Epinimbin-A natural nitrification inhibitor. It is a rich source of plant nutrients, growth promoting substances, nitrogen fixers and phosphorous solubilisers which contribute to vigorous growth and high quality yield

### Neemgard

The extracted powder from processed neem leaf, which is used in several herbal cosmetic preparations, medicated herbal tea., Health, and hygiene products.Neemgard effectively controls various fungus and pests during storage of seeds. This is also used at nursery level for sowing of seeds and it controls seeds from attack of various fungus and viruses particularly nematodes.

### NeemHit Petspray

NeemHit Petspray is a formulation containing neem kernel extract containing azadirachtin for prevention parasite attacks by scabies, eczema, and mange organisms in pets. The protective thin layerAzadirachtin coating is firmly attached to the animal's skin and fur and protect it in a totally natural way, from fleas, lice, ticks, mosquito, sandfly, species of midge etc and which also act as an antifeedant, deterring and repelling all young fleas and killing flea eggs or larvae. NeemHit petspray totally natural product for the skin, 100% ecologically safe, environmentally friendly with antibecterial, antifungal, antiviral properties, that has value-added benefits, like an active that arrests any inflammation caused by a prior bite, as well as help to improve the condition of the skin and hair.

## Chemistry of Ingredients of Neem

Neem plants, as do all other plants, contain several thousands of chemical constituents.Of special interest are the terpenoids are known from different parts of the neem plant. Of its biological constituents the most active and well studied compound is Azadirachtin. However, in most traditional preparations of neem as pesticide or medicine a mixture of neem chemicals are present and provide the active principles. Several different kinds of azadirachtins (A to K) have been isolated, the most abundant of which is Azadirachtin-A.

The neem terpenoids are present in all parts of the plant, in the living tissues. Recently, the site of synthesis and accumulation of the neem chemicals has been identified as secretory cells. Secretory cells are most abundant in the seed kernels. The secretory cells can be seen with iodine solution. Besides the terpenoids, neem also contains more than 20 sulphurous compounds responsible for the characteristic smell of crushed seeds and neem oil.

## Toxicity

In toxicological studies carried out in the USA and Germany, different neem product were neither mutagenous nor cancerogenic, and they did not produce any skin irritations or organic alternations to mice and rates even at high concentrations. In another Canadian study, Neem was found to be harmless to Aquatic invertebrates and other non-target species:

*"To the best of my knowledge, no plant material with greater activity against abroader spectrum of pest insect species, has yet been found."*

The use of Neem in Bangladesh has been dramatically reduced due to destruction of the trees and emergance of chemical industries. Most of Bangladesh was originally forested with coastal mangroves backed by swamp forests and a broad plain of tropical moist deciduous forest (IUCN, 1987). After deforestation the Asian Development Bank (The World Bank) funded aforestation programme selected exotic species from abroad like Eucalytus sp., Dalibergia Sisso, Leucaena leucocephla, Swiietnia macrophulla, and Leucocedha switternia. which grow faster than local natural tress depleting soil and environment. Unfortunately, Neem was not included in the list. About one hundred year ago Neem plant was imported to Africa, where other plants die because of locus/insects attacks but only Neem flourish with wide branches and leaves. I have seen in Sudan only Neem tree survived, whereas other trees perished. Neem tree can easily grow in sandy soils of coastal area of Bangladesh.

Leaf, bark, seed and all part of Neem tree contain useful substances that can be taken as tee, oil and prepared medicine remedy dust allergy, fever, skin diseases, rheumatism etc (Roemmming, 1999, Natur). Professor Heinz Rembold, of famous Max-Planck Institute of Biochemistry, Germany found any side effects of the use of Neem on human and soils do not contain any hazardous substance as a residue after being used as pesticides. In ancient Sanskrit literature (1500 BC) Neem regarded as life saving and disease preventive plant. It belongs to the family Meliaceae.

## Neem Seed

Neem has more than 60 valuable compounds. Over 2000 years that Neem based pesticides have been used in India, many complex processes were developed to make them available for specific use (CSE, India, 2000). Azadirachtin A is the most important biopesticide. There are four important components that make Neem seed the king of the bio-pesticide that prevent and kill more than 400 harmful insects, nematode, fungus, bacteria and virus:

- Azadirachtin
- Meliantriol
- Salannin
- Nimbin.

When the insects eat treated plant, the Neem substances drastically influence the important aspect of the life cycle that prevent the insect further reproducing. A rapid interruption occurs in metabolism, growth and hormone system of the insect that the insects can further reproduce. Another effective advantage is that the insects do not develop resistance.

A neem plant gives about 20-30 kg of seeds. The crushed seeds in water can be used as a powerful pesticide. Since the method has to be repeated, a farmer requires two to three trees for his plants.

Other pesticidal activity includes of need include: (1) The formation of chitin (exoskeleton) is also inhibited. (2) Mating as well as sexual communication is disrupted.(3) Larvae and adults of insects are repelled. (4) Adults are sterilised. (5) larvae and adults are poisoned.

## Neem Oil

Neem oil is an important export item from India Cold pressed oil obtained by traditional grinding method contains most of the useful and biological active ingredients. The oil contains Glyceriden, linol acid, limonoide etc. One kilogram dry seed produces about 100 ml oil. The oil constitutes antiseptic and many medical properties that can be use as ointment, furniture varnish, shampoo and cosmetic articles. The Neem oil has been successfully used by the scientists in the industrial countries to remove different plant diseases.Neem oil also displays numerous remarkably proven medicinal properties also stimulative and antiseptic effect when used for massage of the body. It antiseptic properties has been used to particular advantage in the manufacture of special medicated soaps and tooth pastes, in addition to pharmaceutical preparations like emulsions and ointments. Neem oil has been a important ingredient in soap manufacture and neem oil wood treatment for termite free application.

## Neem Leaf

That dry Neem leaves in rice, lentils keep away insects, fungicides is a traditional wisdom but this is not used any more. Many dangerous pesticides are used for the conservation of food products (Anwar, 1993). Neem Leaf extract shows an extraordinary reaction against dangerous fungus (Aspergillus flavus). It stops producing Aflatoxin (one of the most dangerous cancer producing substance).

## Neem Bark

Neem bark also contains antiseptic properties. Branches are used as tooth brush in the villages. Scientific investigations in Europe have shown that toothpaste prepared from powdered Neem bark has high value for preventive and curative dental treatment. Neem tooth paste is alsobecoming popular in Europe.

## Rediscover Traditional Wisdom

Neem is used for curing skin diseases, muscular pain, nail fungus and many uses that still imperfectly known (i.e. birth control). Unfortunately the use of Neem has decreased dramatically in Bangladesh. Firstly, it is regarded as primitive, socially poor and campaign of chemical industry replaced all traditional uses.

## Mode of Action

Neem acts as a biopesticide at different levels and ways. This is very important since the farmer is used to the knock out effect of chemical pesticides.

Neem does not exhibit this type of effect on pests but affects them in several other ways

## Insect Growth Regulation

It is a very interesting property of neem products and unique in nature, since it works on juvenile hormone. The insect larva feeds when it grows, it sheds the old skin and again starts growing. This particular shedding of old skin is the phenomenon of ecdysis or moulting is governed by an enzyme ecdysone. When the neem components, especially Azadirachtin enter into the body of larvae, the activity of ecdysone is suppressed and the larva fails to moult, remains in the larval stage and ultimately dies. If the concentration of Azadirachtin is not sufficient, the larva manages to enter the pupal stage but dies at this stage and if the concentration is still less the adult emerging from the pupa is 100 % malformed, absolutely sterile without any capacity for reproduction.

## Feeding Deterrent

The most important property of neem is feeding deterrence. When an insect larva sits on the leaf, the larva is hungry and it wants to feed on the leaf. This particular trigger of feeding is given through the maxillary glands give a trigger, peristalsis in the alimentary canal is speeded up, the larva feels hungry and its starts feeding on the surface of the leaf. When the leaf is treated with neem product, because of the presence of azadirachtin, salanin and melandriol there is an anti-peristalitic wave in the alimentary canal and this produces something similar to vomiting sensation in the insect. Because of this sensation the insect does not feed on the neem treated surface. Its ability to swallow is also blocked.

# AGRICULTURE, GLOBALISATION AND BIOTECHNOLOGY

The story of globalisation of ayurveda is also not the story of opening up of a new world of unlimited opportunities as a result of the rise of the herbal products industry worldwide. A certain kind of opportunities has certainly opened up, but by closing down some other possible openings and by changing the very nature of what was and has come to be recognised as ayurvedic medicine. The change is certainly not for the better. Indeed, there is a case for regarding these changes as downgrading of ayurvedic medicine and reducing it to a more rudimentary form of herbal medicine.

With enormous pressures being exerted by the dominant establishment including the pharmaceuticals industry, alternative medical systems have been confined to marketing alternative products.

The real challenge for ayurveda in the global economy lies in defining the parameters and terms of those parts of its knowledge system that are considered adaptable to the market. However, in the scramble to protect markets and knowledge regimes, it is not yet understood that there is a deeper colonisation being played out in the edging out of alternative world-views inherent in these medical systems (M. Banerjee, EPW, January 3, 2004).

The so-called G21 grouping, in Cancun, which represents more than half the world's population and some two-thirds of its farmers, is united by a common commitment to getting the West to unwind subsidies running at nearly $1 billion a day.

"These are the pressures and blackmail we were going through. They are talking about trade liberalisation and that is their mantra. But then in the areas where they do not have an advantage, like agriculture, they practise protectionism. They have double standards, and the people in those countries need to question their government." (Comment from Uganda, Cancun, 2003). The USA and the rich European countries are trying to create a scenario about which we have read in John Steinbeck's novel The Grapes of Wrath. As this is being written the third draft agreement has also faced opposition from the developing countries. "Developing countries have rejected the EU's anti-development agenda. EU member states such as Britain must now start listening to the emerging opposition of developing countries and address their concerns" (NGOs, Reuters, 2003).

Bangladesh's focus on quota-and duty-free access of LDCs' products to the markets in the developed world appeared to centre around non-agricultural products. Then the issue of movement of semi-skilled workers, which Bangladesh raised so ardently as the coordinator of the LDCs, failed to evoke the desired response (Daily Star, Sept., 2003)

A United Nations report said Friday (30.06.06) that globalisation has failed to close glaring inequalities between rich and poor nations and called for developing countries to be given more space to build up their national economies. The UN's '2006 World Economic and Social Survey' said that inequalities at the global level had grown sharply in recent decades, once fast growing China and India-which between them account for one-third of the world's population-were left out of the equation. It argued that poorer countries need to be given more opportunity to diversify their commodity-based economies to make them less vulnerable to fluctuations in world prices and shocks in international financial markets.

Developing countries should be allowed to implement support measures for nascent export industries, and be granted more special and differential treatment under World Trade Organisation agreements, the report said. Support measures for exports are normally frowned upon in the WTO (Agence France-Presse. Geneva, 2.07.06).

Some Western envoys had expressed skepticism that the G21 would survive long because countries such as Brazil and Argentina, efficient farm goods exporters, appeared to have little in common with India, a protectionist nation of 650 million poor farmers.

Professor Stiglitz, the prime voice against globalisation, has noted in his lectures in Bangladesh and elsewhere in the region that "No agreement is better than a bad agreement." It will be incumbent on the Bangladesh delegation and that of the like-minded LDCs to appreciate and act according to Professor's Stiglitz's views on globalisation.

The late 19th and early 20th centuries also saw a globalisation in economic markets, although not as pervasive and widespread as is the case at the present time. Some claim that globalisation has originated from the dynamics of the phenomenal technological advances, while others, particularly anti-globalisation lobbies, contend that big multinational and transnational companies are the responsible parties for both initiating and carrying the process forward.

The origins of globalisation lie in the political decision by the developed countries, led by the USA with active support from the UK. The purpose, it is suggested, was to develop such an international financial architecture and a trade regime as would mobilise savings from around the world to serve the economic interests of the developed countries, particularly the USA.

Thus, those countries insist that the developing countries open up their markets; but they themselves do not walk the promises made and even the agreements granting preferential market access to imports, particularly to non-agricultural imports, from the developing countries.

The level of agricultural subsidy in the USA and the rich EU countries runs at such a high level as US$1 billion a day. On the other hand, they insist that the developing countries do not provide subsidies to agriculture or to any other economic activity.

Moreover, international free movement of labour is not allowed, thereby making the on-going globalisation a partial process, denying the developing countries the one opportunity from which they certainly stand to gain.

The Millennium Developing Goals (MDGs), formulated by the United Nations, focus on issues of deep concern in the developing world. These issues include poverty reduction, promotion of universal primary education, promotion of gender equality and empowerment of women, reduction of child mortality, improvement of maternal health, ensuring environmental sustainability, increasing access to safe drinking water and proper sanitation.

In fact, according to UNDP Human Development Report 2003, 54 countries are now poorer than in 1990. Also, during the same time span, the proportion of people going hungry has increased in 21 countries, and life expectancy at birth has fallen in 34.

The free-market globalisation that is being pushed forward by the international dominant powers through such institutions as the World Bank, the IMF, and the WTO and, also, using bilateral mechanisms, virtually discarding the concept of sustainable development for all practical purposes. The issue of environmental sustainability receives a lot of rhetorical attention, but not much in practical terms. Apathy on the part of the rich countries and lack of human and financial capability as well as of far-sight on the part of most of the developing countries are allowing the process of environmental un-sustainability to continue to accentuate.

The USA, the largest contributor to the global emission of greenhouse gases that are responsible for global warming, has withdrawn from the Kyoto Protocol on control of emission of greenhouse gasses, thereby jeopardising the prospect of the Protocol to come into effect.

## BIODIVERSITY, TRADE AND DEVELOPMENT LINKAGES

The interlinkages between trade, investment, environment, biodiversity, poverty, rural livelihoods and development are multiple and complex, but very crucial in the unequal but globalized world. The world has seen fundamental and many pervasive changes in the last 50 years. The trends toward globalization has been driven in part by the new technologies and in part by reduced barriers to international trade or trade liberalization and investment flows. The result has been a steady increase in the importance of trade and investment in the global economy while the economy quintupled and the world trade grew by a factor of 14 (IIED and DFID, 2002 and UNCTAD, 1999).

On the other hand, it increased global inequality; the benefits of growth have been very unevenly spread and skewed in favour of the developed northern countries. In many cases trade and investment destructed ecology, biodiversity and livelihood of millions of poor particularly in the least developed and developing southern countries (IISD and UNEP, 2000).

Trade liberalization can also increase exploitation of natural resources and exacerbate the associated negative impactson biodiversity. Despite this, a growing number of developing countries look to trade and investment as a central part of their strategies for development and trade considerations are increasingly shaping their economic and development policy.

Biodiversity also has recreational, cultural, spiritual and aesthetic values. Maintaining biodiversity and access to it, while obviously a planetary public good, is crucial for the poor.

The World Health Organization has estimated that 80% of the world's population depends on traditional medicine derived from local plant varieties for their primary health needs.

Wild plants, in field and forest, make a significant contribution to the diet of many poor communities. In many developing countries, poor communities are able to draw at least half of their food from forest products, and consequently have never faced famine.

The emerging global market forces, technological innovation and commercial interest encourage mono-cropping. High technical input and huge investment backed by commercial interest and chief gains in agriculture and other farm level production have destroyed local knowledge and local resources management practices. This process seriously affected the natural resources bases and degraded bioresources. This also dislocated millions of marginal and poor people from their traditional occupation and thus affected their livelihood resulting landlessness, poverty, impoverishment in the development countries. Rapid expansion of shrimp farming and huge investment in shrimp sector by the non-resident rich and power elites in the coastal region of Bangladesh has been one of the classical examples of such unsustainable trade and investment.

1. Destruction of sustainable ecosystem for the finest kitchen of the Industrial Countries
2. Poor farmers losing lands to shrimp farm owners.

The process not only dis-benefited poor in terms of their loss of livelihood and reduced access to natural resources and productive assets, but also eroded their capacity and skills in relation to gaining sustainable livelihood, resources management and conservation of biodiversity. Plantation of exotic tree species in the Madhupur forest in Bangladesh dislocating indigenous people could be an example of such bad investment, where few corrupt people and local power elites played a key role in an ADB supported forestry programme.

1. Plantations (replacing native species) Are Not Forests
2. Forest without Forest Dwellers: Development Program by Asian Development Bank (ADB, The World Bank)

Most of the multinationals and global financial institutions such as the World Bank, IMF, ADB have very often supported the commercial production and high technologies. As a result, a small section of people, mainly big merchants, local agents, few government officials, who control the production, processing and exporting of goods and services have been greatly benefited. On the other hand, the common people and the poor are gradually being marginalized and dis-benefited in the unequal and north dominated trade and investment regime. The process not only dis-benefited poor in terms of their loss of livelihood and reduced access to natural resources and productive assets, but also eroded their capacity and skills in relation to gaining sustainable livelihood, resources management and conservation of biodiversity.

The current WTO rules are too deeply influenced by the powerful trading nations, multinationals and liberalization has dis-benefited the developing countries.Many developing countries have criticized Trade Related Intellectual Property Rights (TRIPS), because it willfavour the developed countries and transnational corporations. TRIPS does not provide any guarantee or safeguard to ensure that the poor share in billions of dollars that may be made from the South's biological resources or the application of traditional knowledge; and most importantly, TRIPS reduces farmer's access and control over agricultural resources including seeds which are essential for their food (BCAS, 2004)

The UNCED Agenda 21 suggested to making trade and environment mutually supportive for achieving sustainable development for the global community. The Agenda 21 stresses on poverty eradication, environmental protection and conservation of natural resources and bio-resources.t he great disappointment was the absence of new benchmark, target or timelines in the areas addressed in the action plan (BRIDGES, September 2002).

# 6

# Sustainable Agriculture and Environment

## INTRODUCTION

The uniqueness of microorganisms and their often unpredictable nature and biosynthetic capabilities, given a specific set of environmental and cultural conditions, has made them likely candidates for solving particularly difficult problems in the life sciences and other fields as well. The various ways in which microorganisms have been used over the past 50 years to advance medical technology, human and animal health, food processing, food safety and quality, genetic engineering, environmental protection, agricultural biotechnology, and more effective treatment of agricultural and municipal wastes provide a most impressive record of achievement. Many of these technological advances would not have been possible using straightforward chemical and physical engineering methods, or if they were, they would not have been practically or economically feasible.

Nevertheless, while microbial technologies have been applied to various agricultural and environmental problems with considerable success in recent years, they have not been widely accepted by the scientific community because it is often difficult to consistently reproduce their beneficial effects. Microorganisms are effective only when they are presented with suitable and optimum conditions for metabolizing their substrates Including available water, oxygen (depending on whether the microorganisms are obligate aerobes or facultative anaerobes), pH and temperature of their environment. Meanwhile, the various types of microbial cultures and inoculants available in the market today have rapidly increased because of these new technologies. Significant achievements are being made in systems where technical guidance is coordinated with the marketing of microbial products. Since microorganisms are useful in eliminating problems associated with the use of chemical fertilizers and pesticides, they are now widely applied in nature farming and organic agriculture.

Environmental pollution, caused by excessive soil erosion and the associated transport of sediment, chemical fertilizers and pesticides to surface and groundwater, and improper treatment of human and animal wastes has caused serious environmental and social problems throughout the world. Often engineers have attempted to solve these problems using established chemical and physical methods. However, they have usually found that such problems cannot be solved without using microbial methods and technologies in coordination with agricultural production.

For many years, soil microbiologists and microbial ecologists have tended to differentiate soil microorganisms as beneficial or harmful according to their functions and how they affect soil quality, plant growth and yield, and plant health. Beneficial microorganisms are those that can fix atmospheric nitrogen, decompose organic wastes and residues, detoxify pesticides, suppress plant diseases and soil-borne pathogens, enhance nutrient cycling, and produce bioactive compounds such as vitamins, hormones and enzymes that stimulate plant growth. Harmful microorganisms are those that can induce plant diseases, stimulate soil-borne pathogens, immobilize nutrients, and produce toxic and putrescent substances that adversely affect plant growth and health.

A more specific classification of beneficial microorganisms has been suggested by Higa (1991; 1994; 1995) which he refer to as "Effective Microorganisms" or EM. This report presents some new perspectives on the role and application of beneficial microorganism, including EM, as microbial inoculants for shifting the soil microbiological equilibrium in ways that can improve soil quality, enhance crop production and protection, conserve natural resources, and ultimately create a more sustainable agriculture and environment The report also discusses strategies on how beneficial microorganisms, including EM. can be more effective after inoculation into soils.

## INTEGRATING SEASONAL CLIMATE PREDICTION AND AGRICULTURAL MODELS

There is a strong history of integrated climate–crop models to estimate the impacts of climate change scenarios, and adaptive breeding and land use strategies. Advances in climate prediction at a seasonal lead-time have stimulated substantial interest in forecasting crop yields as a means of improving farm management and policy-level interventions in a manner that reduces risk and enhances livelihoods and food security, particularly in the marginal, rainfed, tropical cropping regions that the green revolution largely bypassed.

Farmers experience climate change not as a shifting mean but as climatic variations. Climatic variability at all of its time-scales is, therefore, a current challenge to development. Furthermore, developing flexible, proactive strategies for managing year-to-year climate variations within farming communities and the institutions that interface with them, using advance climate information, is arguably the most concrete step that the agricultural enterprise can take to build resilience to long-term changes in the global climate system. These two realizations appear to be bringing the historically distinct global change adaptation and agricultural development agendas closer together, particularly in Africa.

Several research groups are seeking to advance methodology to integrate crop simulation models with dynamic climate forecast models for application at a range of spatial scales. A range of promising approaches has emerged (Hansen *et al.* submitted*b*) that may have relevance to management of climate variability and application to climate change projections. In this paper, we seek to put such work in context by discussing how integrated climate–crop models operating at a seasonal lead-time might contribute to efforts to address climate-related constraints to agricultural development, with some emphasis on rainfed agriculture in Africa. Section 2 summarizes the role of climatic risk and potential role of seasonal climate prediction in agriculture. Section 3 relates integrated climate–crop modelling to constraints that have been identified to effective use of seasonal forecasts within agriculture. The three subsequent sections propose three areas in which integrated modelling might contribute to more effective use of seasonal climate forecasts for the benefit of smallholder agriculture in the developing countries.

## CLIMATE RISK AND AGRICULTURAL DEVELOPMENT

Climate exerts a profound influence on the lives of poor rural populations who depend on agriculture for livelihood and sustenance, who are unprotected against climate-related diseases, who lack secure access to water and food, and who are vulnerable to hydrometeorological hazard. The year-to-year variability of rainfall is a significant constraint to the sustainability of rainfed farming systems in poorer countries of the tropics. This argument is developed more fully in Hansen *et al.* submitted*a*. Climatic extremes, such as drought and flooding, have a direct and often persistent impact on farmers' assets and livelihoods. Perhaps, more important (Elbers *et al.* 2003), the uncertainty associated with climate variability creates a moving target for management, and is a disincentive to sustainable resource management, and to the intensification and adoption of innovation that is necessary for secure livelihoods and long-term rural prosperity.... Variability by itself is not necessarily welfare-decreasing if it is anticipated and acted upon.

Surprise, however, has adverse consequences since the optimal *ex post* and *ex ante* choices rarely coincide Rainfall variability tends to be the dominant source of livelihood risk in smallholder rainfed agriculture, particularly in dryer environments. In the face of climatic uncertainty, risk-averse farmers employ conservative strategies, including avoidance of improved technology, under-use of fertilizers and shifting from productive to non-productive liquid assets. In sub-Sahara Africa, where soil nutrient depletion is now recognized as a root cause of declining per capita food production and a critical constraint to sustainable livelihoods, the apparent effect of climatic uncertainty on investment in soils is a particular concern. These *ex ante* strategies, designed to buffer against climatic extremes, substantially reduce average income and marginal productivity of assets, and do so disproportionately for the relatively poor.

Despite the tremendous concern about climate change, climate variability has not received as much attention as other development issues, in part because it has been considered part of the environmental baseline that is not amenable to intervention. Climate variations in the months ahead no longer need to be accepted as a total unknown.

The chaotic nature of the atmosphere restricts deterministic prediction of weather events to lead-times of several days. However, year-to-year variations in the atmosphere are influenced by interactions between the atmosphere and the more slowly varying ocean and land surfaces, such as those associated with the El Nino-Southern Oscillation (ENSO) in the tropical Pacific.

Improvements in our understanding of interactions between the atmosphere and its underlying surfaces, advances in modelling the global climate system, and substantial investment in monitoring the tropical oceans now provide a degree of predictability of climate fluctuations at a seasonal (i.e. several months) lead-time in many parts of the world. Goddard *et al.* (2001) provide a useful, recent review of the scientific basis and methodology of seasonal climate prediction. Where the necessary conditions are in place or can be put into place, seasonal climate prediction offers an under-exploited opportunity to manage climate variability: to respond proactively to adverse conditions and exploit favourable conditions.

## INTEGRATED CLIMATE-CROP MODELLING AND THE CLIMATE RISK PROBLEM

The potential for seasonal forecasts to reduce the adverse impacts of climate variability and enhance rural livelihoods has motivated a number of pilot studies targeting smallholder farmers in developing countries. Several pilot studies with African farmers have demonstrated a high level of interest and have identified a range of promising livelihood management responses.

They have also identified several obstacles to using seasonal forecasts effectively to reduce the adverse impacts of climate variability and enhance rural livelihoods. I highlight three for which integrated climate–crop modelling has a potential role. The first is a mismatch between farmers' needs and the scale, relevance and transparency of available forecasts.

The second is the cost, risk and learning time farmers face during adaptation and adoption. Third, propose that failure to obtain adequate institutional, financial and political support is also a key constraint. Farm-scale studies have identified several resource constraints that limit responses to advance climate information, but that might be amenable to intervention if additional institutional actors (e.g. suppliers of production inputs and credit) are engaged. Scaling up beyond plot studies will require substantial investment on the part of national agricultural and meteorological services. Providing comprehensive institutional support, including provision of relevant climate information, appropriate technical guidance and intervention in financial services and production inputs (Hansen *et al.* submitted*a*) will require the involvement of a broader range of institutions that have typically been involved in pilot studies.

Several research groups who are working on applications of seasonal climate forecasts have invested heavily in the use of crop simulation models. Although, the use of weather data from analogue years associated with, e.g. ENSO phases, has been the standard approach, interest in coupling crop simulation with dynamic climate forecast models has increased substantially in recent years. A mismatch between the scale of dynamic climate models and crop simulation models complicates the task.

Crop growth depends more on the distribution of weather within a season than on the season averages that forecasters typically provide. Crop simulation models, generally, assume the small-scale of a homogenous plot, and simulate dynamic interactions between weather, management and crop growth and development on a daily time-step. On the other hand, the spatial resolution of the current generation of general circulation models (GCMs) is on the order of

10 000 $km^2$. Although, GCMs simulate the atmosphere on a sub-daily time-step, the spatial averaging within grid cells distorts day-to-day variability, with potentially serious consequences for simulated crop responses. Fortunately, recent and ongoing research has given rise to several promising approaches to translating seasonal climate forecasts into probabilistic predictions of crop response (reviewed in Hansen *et al.* submitted*b*).

It is clear that applications of seasonal climate forecasts to manage risk are concerned with impacts on, e.g. crops and not climate *per se*. Yet, the most critical opportunities and challenges have more to do with institutional support than technology. If that is the case, then what contribution can agricultural models integrated with seasonal climate prediction models make? I propose three promising contributions: (i) translating seasonal climate forecasts into useful information; (ii) *ex ante* assessment of potential benefits and (iii) fostering and guiding application.

## TRANSLATING CLIMATE FORECASTS INTO USEFUL INFORMATION

Hammer (2000) argued that seasonal climate forecasts have no intrinsic value. First, value comes from improved decisions not information. Second, it is not climatic means but impacts within the system being managed that are relevant to decisions. If climate forecasts are to have value, they must modify decision makers' expectations of the production or economic outcomes that relate to goals, within the system being managed (Luseno *et al.* 2003). Crop models integrated with seasonal climate forecasts provide a means of translating forecasts of seasonal climate anomalies into forecasts of production impacts. The nature of the decision determines the appropriate spatial scale and lead-time of crop forecast information.

### Farm Level Applications

Producers of annual crops must routinely make a range of critical production and livelihood decisions prior to planting that interact with climate, but whose outcome is not realized until harvest several months later. Where predictability is sufficient, seasonal forecasts may provide probabilistic information about crop yields with sufficient lead-time to influence pre-planting decisions.

A mismatch between the content, scale, format and lead-time of available operational climate forecasts, and the information that farmers need for such decisions, is the most often cited constraint to effective use. It is also the constraint that is most within the control of climate information providers.

Although, farmers' information requirements are somewhat context-specific, a few generalizations emerge from experience in a range of contexts. Farmers need information that: (i) can be interpreted at a field scale; (ii) includes information about timing beyond three-month climatic means; (iii) is explicit about accuracy; and (iv) is expressed in terms of impacts and management implications within the agricultural systems that they manage.

In contrast, operational climate forecasts are typically expressed as three month climatic means averaged over large areas. When accuracy is communicated, it is typically expressed as shifts in the probability of experiencing outcomes within each climatological tercile, with no information about how the probabilities are derived.

The widespread awareness associated with the 1997 El Nino event stimulated a great deal of interest and debate about the potential value of climate prediction to poor farmers in developing countries. Some expressed doubts about the feasibility of meeting farmers' information requirements. Is predictability of climate and crop response at the farm scale sufficient to be useful? If not, will responding to forecasts that could be wrong expose farmers to unacceptable risk? Barrett (1998) expressed concern that the level of predictability of crop yields at a farm scale is likely to be inadequate for use by risk-averse smallholder farmers. First, fine-scale spatial variability of rainfall implies that seasonal rainfall predictability is limited to aggregate spatial scales. Second, because crop yield is not a simple function of seasonal total rainfall, the accumulation of errors going from seasonal climatic predictors (e.g. sea surface temperatures), to local seasonal means, to crop response implies that predictions of impacts such as crop response will be less accurate than predictions of climatic means. Barrett called for greater investment in methods for translating climate forecasts into forecasts of relevant farm-level impacts.

Averaging in either time or space tends to increase prediction skill by reducing the random 'noise' component of variability in weather. As Gong *et al.* (2003) demonstrate, the skill of GCM-based seasonal climate forecasts tends to increase with increasing spatial aggregation relative to a single GCM grid cell in several parts of the world.

For the state of Ceará, in northeast Brazil, they also showed that downscaling to individual single stations reduces prediction skill, but that a substantial proportion of the predictability remains. Although limited experience seems to support the generalization that forecasts that are skilful at an aggregate scale show only moderate decline at individual points, there remains a need to evaluate the influence of scale of aggregation on prediction skill systematically in a range of locations.

The argument that predictability of agricultural impacts is necessarily less than that of climatic means overlooks two considerations. First, information (e.g. antecedent rainfall, stored soil moisture) beyond seasonal climatic anomalies can contribute to predictability of crop response. Second, predicting relevant impacts directly from climatic predictors instead of from predicted climatic means reduces accumulation of errors, and potentially incorporates information about rainfall distribution and other relevant meteorological variables that are embedded in climatic predictors, but lost when converting them into seasonal rainfall totals.

In a paper that arguably stimulated much of the interest in using seasonal forecasts to benefit farmers in Africa, Cane *et al.* (1994) showed that ENSO-related sea surface temperatures in the Pacific were correlated more strongly with maize yields (1970–1993) than with rainfall averaged across Zimbabwe. Dilley (1997) found that maize yields in Oaxaca, Mexico, were correlated more strongly with the Southern Oscillation Index than with local rainfall. K. P. C. Rao, the APSRU model to simulate maize yields for the short rains season at Katumani, Kenya, as a function of observed weather and GCM-based monthly hindcasts disaggregated to daily values using a stochastic weather generator.

Correlations between predictions and observations were higher for simulated yields than for seasonal rainfall totals, although the analysis did not account for crop model error. By integrating GCM-based seasonal rainfall forecasts into an operational regional wheat forecasting system in Queensland, Australia, Hansen *et al.* (2004) showed that reported district yields were more predictable than seasonal rainfall totals. They attributed this in part to the influence of observed antecedent rainfall and resulting soil moisture storage on subsequent yields. I anticipate that methodological advances and empirical evidence will address concerns about the predictability of crop response to climate fluctuations. Yet, given the spatial heterogeneity, multiplicity of yield-reducing factors and paucity of data in rainfed farming systems of much of Africa, realistic, site-specific forecasts of crop yields are likely to be elusive for some time.

## Early Warning, Market and Policy Applications

Advance information about crop production is relevant to decision-makers operating at a range of scales from field to national or regional, whose decisions impact the welfare of farmers. Institutions responsible for responding to drought or food crises and traders and investors in commodity markets, are interested in anticipating production at relatively large spatial scales. Because of the adverse impacts of variability on commodity prices, the use of advance information about production to manage markets for staple foods to stabilize prices is appealing from the standpoint of food security. Although, experience with using climate information to intervene in markets is still limited, economy-wide modelling in Mozambique suggests considerable potential aggregate benefits of market applications of climate forecasts.

In food-insecure regions of Africa, early warning systems based on remote sensing of rainfall and vegetation, and monitoring of local commodity markets, provide early indication of likely food shortfalls. When a crisis requires external food aid, it can take several months to verify shortfalls, mobilize donors, transport food to the region and distribute it to the affected population. As a result, assistance often reaches the affected people after they have already suffered adverse health effects, divested productive resources or migrated away from their farms.

Food aid organizations in Africa are increasingly interested in seasonal climate forecasts because of their potential to increase lead-time. For example, in August 2004, stakeholders concerned with food security early warning and response in East Africa initiated a periodic food security outlook forum, associated with the Greater Horn of Africa Climate Outlook Forum, with the goal of projecting likely food security impacts associated with the climate outlook to inform response.

In contrast to many climate-sensitive decisions at the farm level, market and food security early warning applications can often use forecast information at various points within the growing season.

The combination of increased spatial scale (Gong *et al.* 2003) and usefulness of forecasts at shorter lead-times increases the predictability potentially available for such applications. On the other hand, institutional constraints to responding to food crises are likely to require a higher threshold of forecast certainty than individual livelihood decisions.

Operational crop forecasting systems exist in many regions to support market and policy applications and food security early warning systems. These are often based on monitoring weather and crop conditions during the growing season. In some instances, they incorporate regionally calibrated crop models to simulate growth or water stress up to the current point within the growing season, then sample weather for the remainder of the season from past years to estimate yield forecast uncertainty. Skilful seasonal forecasts provide additional information that can increase the accuracy of within-season production forecasts based on monitoring and simulation alone, particularly early in the growing season. Integrating probabilistic forecasts of food production from integrated climate–crop models into models of household vulnerability has potential to improve aid assessments and targeting at a relatively long lead-time.

## *EX ANTE* ASSESSMENT OF BENEFIT

*Ex ante* impact assessment seeks to assess the potential outcomes of an innovation in advance of its adoption, while *ex post* assessment seeks to assess actual outcomes following adoption. Because the use of seasonal climate forecasts within agriculture is a relatively new innovation, *ex ante* methods are, in most cases, the only way to estimate their benefits.

Pilot studies have compiled some evidence of use of forecasts for farm decisions, but have, generally, not tried to quantify the resulting production or livelihood benefits. Even after farmers learn to use forecast information routinely, *ex post* assessment of benefits would require multiple years due to the stochastic nature of climate variability and forecast responses—longer than typical funding cycles allow. Furthermore, there are few, if any, regions in the developing world where rural communities have had access to operational climate information tailored to their needs for sufficient time to allow *ex post* assessment of use and benefits.

*Ex ante* impact assessment serves two related roles in applied agricultural research: providing evidence of potential benefit of proposed interventions to support mobilization of resources, and providing insight to target interventions where the positive impact is likely to be greatest (Thornton submitted). Resource mobilization has to do with allocation of scarce resources between a particular intervention or institution, and others that compete. Targeting relates to an institution or program's allocation of scarce resources among its own competing activities.

### Confidence and Credibility

Increasing competition for dwindling resources for applied research has driven the international agricultural research community to become proactive about both quantifying the benefits of past activities (i.e. *ex post* impact assessment), and providing evidence of the likely benefits of proposed interventions (i.e. *ex ante* assessment). Likewise, donors increasingly require *ex ante* evidence of benefit before they will invest heavily in new agricultural innovations. The application of seasonal climate prediction is a new innovation that potentially competes for scarce resources with technologies such as crop genetic improvements that have relatively long records of *ex post* evidence of impact.

On the positive side, it shares with integrated natural resource management (INRM) a relatively short history, heavy demands on management and context-specific implementation that prevent robust *ex post* impact evaluation (Barrett 2003). Yet, the broad suite of interventions that fall under INRM are generally well funded within the international agricultural research community. Climate applications face the additional challenge of engaging climate and agricultural institutions that often do not have a history of substantive cooperation.

Although, there are good theoretical reasons to expect that farmers and other agricultural stakeholders should benefit, there are still few well documented *ex post* demonstrations of adoption and benefit in developing countries, relative to many other agricultural development interventions. Some rather fundamental expressions of doubt about the potential utility to farmers in developing countries appear in the literature, citing examples of lack of use and barriers to use of available seasonal forecasts by smallholder farmers.

Hansen (2004) summarizes some of these concerns in a set of questions: is there enough predictability at the farm scale to be useful? Do smallholder farmers have the capacity to respond to climate forecasts? Will climate forecasts that could be wrong expose farmers to unacceptable risk? Can smallholder farmers with limited education understand probabilistic climate forecasts? Is the application of seasonal climate forecasts inherently biased against the poor? With the exception of farmers' ability to understand probabilistic forecasts, well designed, data-driven *ex ante* assessments can help answer these concerns for particular contexts.

Credible evidence of acceptability and benefit to farmers may contribute to mobilizing funds and influencing the agendas of institutional partners. On the other hand, much of the support for forecast applications for agriculture has come from sources that do not traditionally fund agricultural research and development, and that have not yet demanded the level of evidence that traditional agricultural donors typically require.

## Targeting Interventions

The potential to benefit from seasonal climate forecasts lies in the intersection of predictability of relevant components of climate variability, vulnerability (more correctly, sensitivity) to the adverse impacts of climate variability and ability to modify climate-sensitive decisions (Hansen 2002). *Ex ante* impact assessment, using integrated climate–crop–economic modelling, can provide a quantitative basis for selecting priorities for allocating scarce resources among regions, farming systems and interventions. Examples of the use of *ex ante* modelling to target regions or populations seem to be absent from the climate applications literature. There are, however, a few examples of analysis that are relevant to setting priorities for forecast system development and policy.

## FOSTERING AND GUIDING RESPONSES: MODELS FOR MANAGEMENT

### Modelling for Agrotechnology Transfer

Crop simulation was initially seen as a way to integrate knowledge of individual processes, obtain insights into interactions, and use predictive ability to test hypotheses. As their ability to simulate complex genotype × environment × management interactions advanced, crop models were increasingly seen as a way to enhance the transferability of results of traditional agronomic research in the face of environmental heterogeneity.

The advances in food security and rural welfare associated with the green revolution in Asia and Latin America have been difficult to reproduce in marginal rainfed regions. Heterogeneity of the environment has been advanced as one explanation for low rates of adoption of improved technology, and weak returns on investment in extension services in Africa relative to other parts of the developing world. 'An important reason for the failure of earlier programmes was their top-down nature and limited recognition of the wide diversity and heterogeneity of farmers and fields as a prime characteristic of livelihoods and farming systems in (less-favoured areas)'.

Because the production outcomes of management are highly dependent on soils, climate and topography and on farmers' goals, assets, social status and access to markets and services, spatial heterogeneity of these variables creates a need to tailor management to local conditions. Experimental methodology developed in the crop sciences to control for the effects of variability in space and time, is not well suited to the task.

Farming systems research (FSR), advanced as a way to understand the context-specific needs and opportunities of farmers, shares with agronomic experimentation the location specificity problem and need to aggregate to recommendation domains in order to scale up results, with an inevitable tradeoff between relevance and cost. The ten-year international benchmark sites network for agrotechnology transfer (IBSNAT) project was developed under the assumption that appropriate use of crop models could compensate for the limitations of traditional research methods in the face of environmental heterogeneity (IBSNAT 1993). Well validated process-oriented models were seen as relatively rapid and cost-effective means of transferring research results to locations other than where they were developed and tested, and tailoring technology to specific environments without having to replicate field research at every location.

The rationale for using system modelling for agrotechnology transfer in the face of environmental heterogeneity is relevant to the use of seasonal forecasts for climate risk management. Climate variability adds a time dimension to environmental heterogeneity. Regionally adapted and tested crop models allow one to quickly explore the production outcomes of a range of management alternatives under a range of climate scenarios (realized or predicted), and for a range of soil conditions.

System modelling is arguably the only tractable way to do so. Although agronomic trials can be replicated in space, at significant cost, to provide information on management outcomes across space, extending replicated trials over enough years to provide robust results across the range of climate variability is not feasible. Furthermore, adaptive climate risk management is not a packaged technology, but an information input with implications for potentially many aspects of the farming and livelihood system. Because of differences in resource endowment, goals, livelihood strategies and risk tolerance, these applications are likely to be more context-specific than many agronomic technologies.

## Farm Management Decision Support Systems

Although, modelling tools that came out of the IBSNAT project and similar efforts during the 1980s and early 1990s targeted researchers, the goal was to use models as '...the means by which knowledge of systems and their performance is made portable and accessible to users whose livelihood and welfare depend on this performance' (IBSNAT 1993). Yet, the mechanism for transferring the resulting knowledge to farmers was not always clear. Increased on-farm use of PCs in developed countries raised the prospect of developing model-based decision support systems (DSS) to deliver the benefits of modelling to farmers more directly through support of tactical farm management decisions.

There have been several attempts to develop and deliver agricultural simulation models via farmer DSS, primarily in Australia and the USA (McCown 2002). To address the problem of model complexity, most provide simplified user interfaces to crop models, or databases of pre-run simulations. PC-Yield addressed the data availability problem by packaging DSS software with a subscription to daily field-specific weather data updates via the Internet. The FARMSCAPE project provided on-site weather stations and soil characterization for client farmers (Carberry *et al.* 2002). Limited experience suggests that significant early and ongoing stakeholder participation is also a necessary condition for success.

Despite some modest successes, uptake of model-based farm DSS has, generally, been disappointing. A review of crop model applications in developing countries brought a growing awareness to the forefront, that crop simulation models have not fulfilled early expectations as a tool for transferring technology to farmers.

From a review of 14 DSS projects, McCown (2002) provides a thoughtful analysis of reasons for the low level of farmer adoption, and suggests several niches where DSS might play a significant role in farm decision-making. One of the key lessons is that, while farmers might be willing to use a DSS as a tool to assist in low-level, routine tasks, they tend to resist any proxy—human or software—for their own management process.

Two other observations about farm DSS are significant to this discussion. First, several authors have noted instances where interactions and experimentation around model-based DSS have challenged farmers' perceptions, and stimulated learning and the formulation of new heuristics (i.e. rules of thumb). Once the new heuristics are learned, the DSS is no longer needed. Second, there appears to be a trend toward delivering the knowledge embodied in model-based DSS through expert intermediaries, rather than directly to farmers via the software.

## From Decision Support to Discussion Support

The shift in focus away from the traditional notion of DSS toward the use of modelling to stimulate dialogue, highlight options and tradeoffs, and foster co-learning and consensus building lead to the relatively new notion of models as 'discussion support systems'. 'In this 'mutual understanding' relationship, intervention intent shifts from educating and persuading to recognition of and respect for other ways of viewing the world....Intervention emphasis shifts from *prescribing* action to *facilitating* learning in actions' (McCown 2002).

This transition in thinking about the role of agricultural simulation models parallels an ongoing shift within the broader agricultural development community away from a linear model of technology development and transfer, towards increasing recognition that farmers are not passive recipients but innovators who actively experiment and repackage technology to meet their needs and circumstances.

The farm DSS projects that McCown (2002) evaluated are all in developed countries. To what degree can this experience translate to smallholder farmers in developing countries? Limited experience suggests that the prospects are quite good. In a set of workshops with farmers in two villages in Zimbabwe, Carberry *et al.* (2004) used APSIM crop simulations as a basis for discussion about a range of farm management questions.

Despite initial scepticism on the part of the research team and participating farmers, the farmers found the simulation results to be credible and relevant. The simulation model facilitated virtual experimentation, allowing researchers to respond quickly to questions that participating farmers raised about crop response to fertilizer, management of a limited supply of manure and impacts on competing crops of planting delays forced by labour shortage.

The researchers concluded that the model facilitated a high level of farmer–researcher dialogue and debate between farmers, challenged farmers' perceptions in a manner that led to learning, and stimulated interest in on-farm testing of innovations proposed by the farmers. In the context of climate risk management, researchers used the APSIM model effectively with a group of farmers in western Tamil Nadu, India, as a basis for discussing and debating possible management responses to ENSO-based climate forecasts.

Shifts in aggregate crop area statistics were consistent with farmer discussions about the 2002 climate forecast and crop model results. Gadgil *et al.* (2002) responded to a request from farmers by using the PNUTGRO model to evaluate climatic risk as a function of planting date for rainfed groundnut in southwestern Andhra Pradesh, India. The farmers appeared to value and learn from crop model results that supported a later planting date than normally recommended (Gadgil personal communication).

## Model-based Recommendations?

Although, crop models have contributed significantly to discussions about climate applications between researchers and collaborating groups of farmers, the cost of such intensive interactions is high relative to the number of farmers impacted. I am aware of several researchers who are interested in using crop models as a basis for formulating advisories for broad distribution within farming communities. The trend in the USA and Australia towards use of crop models by expert intermediaries who advise farmers (McCown 2002) seems to support such an approach.

Just as DSS have, in come cases, prompted learning and formulation of new heuristics that eliminated further need for the DSS among farmers (McCown 2002), it seems likely that crop models integrated with seasonal forecasts may serve more as a vehicle researchers and farmer advisors to learn principles of climate risk management through accelerated experience and experimentation that they can then transfer through their interaction with farmers.

## THE CONCEPT OF EFFECTIVE MICROORGANISMS: THEIR ROLE AND APPLICATION

The concept of effective microorganisms (EM) was developed by Professor Teruo Higa, University of the Ryukyus, Okinawa, Japan. EM consists of mixed cultures of beneficial an naturally-occurring microorganisms that can be applied as inoculants to increase the microbial diversity of soils and plant. Research has shown that the inoculation of EM cultures to the soil/plant ecosystem can improve soil quality, soil health, and the growth, yield, and quality of crops. EM contains selected species of microorganisms including predominant populations of lactic acid bacteria and yeasts and smaller numbers of photosynthetic bacteria, actinomycetes and other types of organisms. All of these are mutually compatible with one another and can coexist in liquid culture.

EM is not a substitute for other management practices. It is, however, an added dimension for optimizing our best soil and crop management practices such as crop rotations, use of organic amendments, conservation tillage, crop residue recycling, and biocontrol of pests. If used properly, EM can significantly enhance the beneficial effects of these practices.

Throughout the discussion which follows, we will use the term "beneficial microorganisms" In a general way to designate a large group of often unknown or ill-defined microorganisms that interact favourably in soils and with plants to render beneficial effects which are sometimes difficult to predict. We use the term "effective microorganisms" or EM to denote specific mixed cultures of known, beneficial microorganisms that are being used effectively as microbial inoculants.

## UTILIZATION OF BENEFICIAL MICROORGANISMS IN AGRICULTURE

### What Constitutes an Ideal Agricultural System?

Conceptual design is important in developing new technologies for utilizing beneficial and effective microorganisms for a more sustainable agriculture and environment. The basis of a conceptual design is imply to first conceive an ideal or model and then to devise a strategy and method for achieving the reality. However it is necessary to carefully coordinate the materials, the environment, and the technologies constituting the method.

Moreover one should adopt a philosophical attitude in applying microbial technologies to agricultural production and conservation systems.

There are many opinions on what an ideal agricultural system is. Many would agree that such an idealized system should produce food on a long-term sustainable basis. Many would also insist that it should maintain and improve human health, be economically and spiritually beneficial to both producers and consumers, actively preserve and protect the environment, be self-contained and regenerative, and produce enough food for an increasing world population.

## Efficient Utilization and Recycling of Energy

Agricultural production begins with the process of photosynthesis by green plants which requires solar energy, water, and carbon dioxide. It occurs through the plants ability to utilize solar energy in "fixing" atmospheric carbon into carbohydrates. The energy obtained is used for further biosynthesis in the plant, including essential amino acids and proteins. The materials used for agricultural production are abundantly available with little initial cost. However, when it is observed as an economic activity, the fixation of carbon by photosynthesis has an extremely low efficiency mainly because of the low utilization rate of solar energy by green plants. Therefore, an integrated approach is needed to increase the level of solar energy utilization by plants so that greater amounts of atmospheric carbon can be converted into useful substrates.

Although the potential utilization rate of solar energy by plants has been estimated theoretically at between 10 and 20%, the actual utilization rate is less than 1%. Even the utilization rate of C4 plants, such as sugar cane whose photosynthetic efficiency is very high, barely exceeds 6 or 7% during the maximum growth period. The utilization rate is normally less than 3% even for optimum crop yields.

Past studies have shown that photosynthetic efficiency of the chloroplasts of host crop plants cannot be increased much further; this means that their biomass production has reached a maximum level. Therefore, the best opportunity for increasing biomass production is to somehow utilize the visible light, which chloroplasts cannot presently use, and the infrared radiation; together, these comprise about 80% of the total solar energy. Also, we must explore ways of recycling organic energy contained in plant and animal residues through direct utilization of organic molecules by plants.

Thus, it is difficult to exceed the existing limits of crop production unless the efficiency of utilizing solar energy is increased, and the energy contained in existing organic molecules (amino acids, peptides and carbohydrates) is utilized either directly or indirectly by the plant. This approach could help to solve the problems of environmental pollution and degradation caused by the misuse and excessive application of chemical fertilizers and pesticides to soils. Therefore, new technologies that can enhance the economic-viability of farming systems with little or no use of chemical fertilizers and pesticides are urgently needed and should be a high priority of agricultural research both now and in the immediate future.

### Preservation of Natural Resources and the Environment

The excessive erosion of topsoil from farmland caused by intensive tillage and row-crop production has caused extensive soil degradation and also contributed to the pollution of both surface and groundwater. Organic wastes from animal production, agricultural and marine processing industries, and municipal wastes (i.e., sewage and garbage), have become major sources of environmental pollution in both developed and developing countries. Furthermore, the production of methane from paddy fields and ruminant animals and of carbon dioxide from the burning of fossil fuels, land clearing and organic matter decomposition have been linked to global warming as "greenhouse gases".

Chemical-based, conventional systems of agricultural production have created many sources of pollution that, either directly or indirectly, can contribute to degradation of the environment and destruction of our natural resource base. This situation would change significantly if these pollutants could be utilized in agricultural production as sources of energy.

Therefore, it is necessary that future agricultural technologies be compatible with the global ecosystem and with solutions to such problems in areas different from those of conventional agricultural technologies. An area that appears to hold the greatest promise for technological advances in crop production, crop protection, and natural resource conservation is that of beneficial and effective microorganisms applied as soil, plant and environmental inoculants (Higa, 1995).

## BENEFICIAL AND EFFECTIVE MICROORGANISMS FOR A SUSTAINABLE AGRICULTURE

Agriculture in a broad sense, is not an enterprise which leaves everything to nature without intervention. Rather it is a human activity in which the farmer attempts to integrate certain agroecological factors and production inputs for optimum crop and livestock production. Thus, it is reasonable to assume that farmers should be interested in ways and means of controlling beneficial soil microorganisms as an important component of the agricultural environment. Nevertheless, this idea has often been rejected by naturalists and proponents of nature farming and organic agriculture. They argue that beneficial soil microorganisms will increase naturally when organic amendments are applied to soils as carbon, energy and nutrient sources. This indeed may be true where an abundance of organic materials are readily available for recycling which often occurs in small-scale farming. However, in most cases, soil microorganisms, beneficial or harmful, have often been controlled advantageously when crops in various agroecological zones are grown and cultivated in proper sequence (i.e., crop rotations) and without the use of pesticides. This would explain why scientists have long been interested in the use of beneficial microorganisms as soil and plant inoculants to shift the microbiological equilibrium in a way that enhances soil quality and the yield and quality of crops.

Most would agree that a basic rule of agriculture is to ensure that specific crops are grown according to their agroclimatic and agroecological requirements. However, in many cases the agricultural economy is based on market forces that demand a stable supply of food, and thus, it becomes necessary to use farmland to its full productive potential throughout the year.

The purpose of crop breeding is to improve crop production, crop protection, and crop quality. Improved crop cultivars along with improved cultural and management practices have made it possible to grow a wide variety of agricultural and horticultural crops in areas where it once would not have been culturally or economically feasible. The cultivation of these crops in such diverse environments has contributed significantly to a stable food supply in many countries. However, it is somewhat ironic that new crop cultures are almost never selected with consideration of their nutritional quality or bioavailability after ingestion (Hornick, 1992).

As will be discussed later, crop growth and development are closely related to the nature of the soil microflora, especially those in close proximity to plant roots, i.e., the rhizosphere. Thus, it will be difficult to overcome the limitations of conventional agricultural technologies without controlling soil microorganisms. This particular tenet is further reinforced because the evolution of most forms of life on earth and their environments are sustained by microorganisms. Most biological activities are influenced by the state of these invisible, minuscule units of life. Therefore, to significantly increase food production, it is essential to develop crop cultivars with improved genetic capabilities (i.e., greater yield potential, disease resistance, and nutritional quality) and with a higher level of environmental competitiveness, particularly under stress conditions (i.e., low rainfall, high temperatures, nutrient deficiencies, and agressive weed growth).

To enhance the concept of controlling and utilizing beneficial microorganisms for crop production and protection, one must harmoniously integrate the essential components for plant growth and yield including light (intensity, photoperiodicity and quality), carbon dioxide, water, nutrients (organic-inorganic) soil type, and the soil microflora. Because of these vital interrelationships, it is possible to envision a new technology and a more energy-efficient system of biological production.

Low agricultural production efficiency is closely related to a poor coordination of energy conversion which, in turn, is influenced by crop physiological factors, the environment, and other biological factors including soil microorganisms. The soil and rhizosphere microflora can accelerate the growth of plants and enhance their resistance to disease and harmful insects by producing bioactive substances. These microorganisms maintain the growth environment of plants, and may have secondary effects on crop quality. A wide range of results are possible depending on their predominance and activities at any one time. Nevertheless, there is a growing consensus that it is possible to attain maximum economic crop yields of high quality, at higher net returns, without the application of chemical fertilizers and pesticides. Until recently, this was not thought to be a very likely possibility using conventional agricultural methods. However, it is important to recognize that the best soil and crop management practices to achieve a more sustainable agriculture will also enhance the growth, numbers and activities of beneficial soil microorganisms that, in turn, can improve the growth, yield and quality of crops.

## CONTROLLING THE SOIL MICROFLORA: PRINCIPLES AND STRATEGIES

### Principles of Natural Ecosystems and the Application of Beneficial and Effective Microorganisms

The misuse and excessive use of chemical fertilizers and pesticides have often adversely affected the environment and created many a) food safety and quality and b) human and animal health problems. Consequently, there has been a growing interest in nature farming and organic agriculture by consumers and environmentalists as possible alternatives to chemical-based, conventional agriculture.

Agricultural systems which conform to the principles of natural ecosystems are now receiving a great deal of attention in both developed and developing countries. A number of books and journals have recently been published which deal with many aspects of natural farming systems. New concepts such as alternative agriculture, sustainable agriculture, soil quality, integrated pest management, integrated nutrient management and even beneficial microorganisms are being explored by the agricultural research establishment. Although these concepts and associated methodologies hold considerable promise, they also have limitations. For example, the main limitation in using microbial inoculants is the problem of reproducibility and lack of consistent results.

Unfortunately certain microbial cultures have been promoted by their suppliers as being effective for controlling a wide range of soil-borne plant diseases when in fact they were effective only on specific pathogens under very specific conditions. Some suppliers have suggested that their particular microbial inoculant is akin to a pesticide that would suppress the general soil microbial population while increasing the population of a specific beneficial microorganism. Nevertheless, most of the claims for these single-culture microbial inoculants are greatly exaggerated and have not proven to be effective under field conditions. One might speculate that if all of the microbial cultures and inoculants that are available as marketed products were used some degree of success might be achieved because of the increased diversity of the soil microflora and stability that is associated with mixed cultures. While this, of course, is a hypothetical example, the fact remains that there is a greater likelihood of controlling the soil microflora by introducing mixed, compatible cultures rather than single pure cultures.

Even so, the use of mixed cultures in this approach has been criticized because it is difficult to demonstrate conclusively which microorganisms are responsible for the observed effects, how the introduced microorganisms interact with the indigenous species, and how these new associations affect the soil/plant environment. Thus, the use of mixed cultures of beneficial microorganisms as soil inoculants to enhance the growth, health, yield, and quality of crops has not gained widespread acceptance by the agricultural research establishment because conclusive scientific proof is often lacking.

The use of mixed cultures of beneficial microorganisms as soil inoculants is based on the principles of natural ecosystems which are sustained by their constituents; that is, by the quality and quantity of their inhabitants and specific ecological parameters, i.e., the greater the diversity and number of the inhabitants, the higher the order of their interaction and the more stable the ecosystem. The mixed culture approach is simply an effort to apply these principles to natural systems such as agricultural soils, and to shift the microbiological equilibrium in favour of increased plant growth, production and protection.

It is important to recognize that soils can vary tremendously as to their types and numbers of microorganisms. These can be both beneficial and harmful to plants and often the predominance of either one depends on the cultural and management practices that are applied. It should also be emphasized that most fertile and productive soils have a high content of organic matter and, generally, have large, populations of highly diverse microorganisms (i.e., both species and genetic diversity). Such soils will also usually have a wide ratio of beneficial to harmful microorganisms.

## Controlling the Soil Microflora for Optimum Crop Production and Protection

The idea of controlling and manipulating the soil microflora through the use of inoculants organic amendments and cultural and management practices to create a more favorable soil microbiological environment for optimum crop production and protection is not new. For almost a century, microbiologists have known that organic wastes and residues, including animal manures, crop residues, green manures, municipal wastes (both raw and composted), contain their own indigenous populations of microorganisms often with broad physiological capabilities.

It is also known that when such organic wastes and residues are applied to soils many of these introduced microorganisms can function as biocontrol agents by controlling or suppressing soil-borne plant pathogens through their competitive and antagonistic activities. While this has been the theoretical basis for controlling the soil microflora, in actual practice the results have been unpredictable and inconsistent, and the role of specific microorganisms has not been well-defined.

For, many years microbiologists have tried to culture beneficial microorganisms for use as soil inoculants to overcome the harmful effects of phytopathogenic organisms, including bacteria, fungi and nematodes. Such attempts have usually involved single applications of pure cultures of microorganisms which have been largely unsuccessful for several reasons. First, it is necessary to thoroughly understand the individual growth and survival characteristics of each particular beneficial microorganism, including their nutritional and environmental requirements. Second, we must understand their ecological relationships and interactions with other microorganisms, including their ability to coexist in mixed cultures and after application to soils. There are other problems and constraints that have been major obstacles to controlling the microflora of agricultural soils.

First and foremost is the large number of types of microorganisms that are present at any one time, their wide range of physiological capabilities, and the dramatic fluctuations in their populations that can result from man's cultural and management practices applied to a particular farming system. The diversity of the total soil microflora depends on the nature of the soil environment and those factors which affect the growth and activity of each individual organism including temperature, light, aeration, nutrients, organic matter, pH and water. While there are many microorganisms that respond positively to these factors, or a combination thereof, there are many that do not. Microbiologists have actually studied relatively few of the microorganisms that exist in most agricultural soil, mainly because we don't know how to culture them; i.e., we know very little about their growth, nutritional, and ecological requirements.

The "diversity" and "population" factors associated with the soil microflora have discouraged scientists from conducting research to develop control strategies. Many believe that, even when beneficial microorganisms are cultured and inoculated into soils, their number is relatively small compared with the indigenous soil inhabitants, and they would likely be rapidly overwhelmed by the established soil microflora. Consequently, many would argue that even if the application of beneficial microorganisms is successful under limited conditions (e.g., in the laboratory) it would be virtually impossible to achieve the same success under actual field conditions. Such thinking still exists today, and serves as a principle constraint to the concept of controlling the soil microflora.

It is noteworthy that most of the microorganisms encountered in any particular soil are harmless to plants with only a relatively few that function as plant pathogens or potential pathogens. Harmful microorganisms become dominant if conditions develop that are favorable to their growth, activity and reproduction. Under such conditions, soil-borne pathogens (e.g., fungal pathogens) can rapidly increase their populations with devastating effects on the crop. If these conditions change, the pathogen population declines just as rapidly to its original state. Conventional farming systems that tend toward the consecutive planting of the same crop (i.e., monoculture) necessitate the heavy use of chemical fertilizers and pesticides.

This, in turn, generally increases the probability that harmful, disease-producing, plant pathogenic microorganisms will become more dominant in agricultural soils. Chemical-based conventional farming methods are not unlike symptomatic therapy. Examples of this are applying fertilizers when crops show symptoms of nutrient-deficiencies, and applying pesticides whenever crops are attacked by insects and diseases. In efforts to control the soil microflora some scientists feel that the introduction of beneficial microorganisms should follow a symptomatic approach. However, we do not agree. The actual soil conditions that prevail at any point in time may be most unfavourable to the growth and establishment of laboratory-cultured, beneficial microorganisms. To facilitate their establishment, it may require that the farmer make certain changes in his cultural and management practices to induce conditions that will (a) allow the growth and survival of the inoculated microorganisms and (b) suppress the growth and activity of the indigenous plant pathogenic microorganisms.

An example of the importance of controlling the soil microflora and how certain cultural and management practices can facilitate such control is useful here. Vegetable cultivars are often selected on their ability to grow and produce over a wide range of temperatures. Under cool, temperate conditions there are generally few pest and disease problems. However, with the onset of hot weather, there is a concomitant increase in the incidence of diseases and insects making it rather difficult to obtain acceptable yields without applying pesticides. With higher temperatures, the total soil microbial population increases as does certain plant pathogens such as Fusarium, which is one of the main putrefactive, fungal pathogens in soil. The incidence and destructive activity of this pathogen can be greatly minimized by adopting reduced tillage methods and by shading techniques to keep the soil cool during hot weather. Another approach is to inoculate the soil with beneficial, antagonistic, antibiotic-producing microorganisms such as actinomycetes and certain fungi.

## Application of Beneficial and Effective Microorganisms: A New Dimension

Many microbiologists believe that the total number of soil microorganisms can be increased by applying organic amendments to the soil. This is generally true because most soil microorganisms are heterotrophic, i.e., they require complex organic molecules of carbon and nitrogen for metabolism and biosynthesis. Whether the regular addition of organic wastes and residues will greatly increase the number of beneficial soil microorganisms in a short period of time is questionable. However, we do know that heavy applications of organic materials, such as seaweed, fish meal, and chitin from crushed crab shells, not only helps to balance the micronutrient content of a soil but also increases the population of beneficial antibiotic-producing actinomycetes. This changes the soil to a disease-suppressive condition within a relatively short period.

The probability that a particular beneficial microorganism will become predominant, even with organic farming or nature farming methods, will depend on the ecosystem and environmental conditions. It can take several hundred years for various species of higher and lower plants to interact and develop into a definable and stable ecosystem. Even if the population of a specific microorganism is increased through cultural and management practices, whether it will be beneficial to plants is another question. Thus, the likelihood of a beneficial, plant-associated microorganism becoming predominant under conservation-based farming systems is virtually impossible to predict. Moreover, it is very unlikely that the population of useful anaerobic microorganisms, which usually comprise only a small part of the soil microflora, would increase significantly even under natural farming conditions.

This information then emphasizes the need to develop methods for isolating and selecting different microorganisms for their beneficial effects on soils and plants. The ultimate goal is to select microorganisms that are physiologically and ecologically compatible with each other and that can be introduced as mixed cultures into soil where their beneficial effects can be realized.

## Application of Beneficial and Effective Microorganisms: Fundamental Considerations

Microorganisms are utilized in agriculture for various purposes; as important components of organic amendments and composts, as legume inoculants for biological nitrogen fixation as a means of suppressing insects and plant diseases to Improve crop quality and yields, and for reduction of labour. All of these are closely related to each other. An important consideration in the application of beneficial microorganisms to soils is the enhancement of their synergistic effects. This is difficult to accomplish if these microorganisms are applied to achieve symptomatic therapy, as in the case of chemical fertilizers and pesticides.

If cultures of beneficial microorganisms are to be effective after inoculation into soil, it is important that their initial populations be at a certain critical threshold level. This helps to ensure that the amount of bioactive substances produced by them will be sufficient to achieve the desired positive effects on crop production and/or crop protection. If these conditions are not met, the introduced microorganisms, no matter how useful they are, will have little if any effect. At present, there are no chemical tests that can predict the probability of a particular soil-inoculated microorganism to achieve a desired result. The most reliable approach is to inoculate the beneficial microorganism into soil as part of a mixed culture, and at a sufficiently high inoculum density to maximize the probability of its adaptation to environmental and ecological conditions.

The application of beneficial microorganisms to soil can help to define the structure and establishment of natural ecosystems. The greater the diversity of the cultivated plants that are grown and the more chemically complex the biomass, the greater the diversity of the soil microflora as to their types, numbers and activities.

The application of a wide range of different organic amendments to soils can also help to ensure a greater microbial diversity. For example, combinations of various crop residues, animal manures, green manures, and municipal wastes applied periodically to soil will provide a higher level of microbial diversity than when only one of these materials is applied. The reason for this is that each of these organic materials has its own unique indigenous microflora which can greatly affect the resident soil microflora after they are applied, at least for a limited period.

## CLASSIFICATION OF SOILS BASED ON THEIR MICROBIOLOGICAL PROPERTIES

Most soils are classified on the basis of their chemical and physical properties; little has been done to classify soils according to their physicochemical and microbiological properties. The reason for this is that a soil's chemical and physical properties are more readily defined and measured than their microbiological properties.

Improved soil quality is usually characterized by increased infiltration; aeration, aggregation and organic matter content and by decreased bulk density, compaction, erosion and crusting. While these are important indicators of potential soil productivity, we must give more attention to soil biological properties because of their important relationship (though poorly understood) to crop production, plant and animal health, environmental quality, and food safety and quality. Research is needed to identify and quantify reliable and predictable biological/ecological indicators of soil quality. Possible indicators might include total species diversity or genetic diversity of beneficial soil microorganisms as well as insects and animals.

The basic concept here is not to classify soils for the study of microorganisms but for farmers to be able to control the soil microflora so that biologically-mediated processes can improve the growth, yield, and quality of crops as well as the tilth, fertility, and productivity of soils. The ultimate objective is to reduce the need for chemical fertilizers and pesticides.

## Functions of Microorganisms: Putrefaction, Fermentation, and Synthesis

Soil microorganisms can be classified into decomposer and synthetic microorganisms. The decomposer microorganisms are subdivided into groups that perform oxidative and fermentative decomposition. The fermentative group is further divided into useful fermentation (simply called fermentation) and harmful fermentation (called putrefaction). The synthetic microorganisms can be sub-divided into groups having the physiological abilities to fix atmospheric nitrogen into amino acids and/or carbon dioxide into simple organic molecules through photosynthesis. Adapted from Higa is a simplified flow chart of organic matter transformations by soil microorganisms that can lead to the development of disease-inducing, disease-suppressive, zymogenic, or synthetic soils.

Fermentation is an anaerobic process by which facultative microorganisms (e.g., yeasts) transform complex organic molecules (e.g., carbohydrates) into simple organic compounds that often can be absorbed directly by plants. Fermentation yields a relatively small amount of energy compared with aerobic decomposition of the same substrate by the same group of microorganisms.

Aerobic decomposition results in complete oxidation of a substrate and the release of large amounts of energy, gas, and heat with carbon dioxide and water as the end products. Putrefaction is the process by which facultative heterotrophic microorganisms decompose proteins anaerobically, yielding malodorous incompletely oxidized, metabolites (e.g., ammonia, mercaptans and indole) that are often toxic to plants and animals.

The term "synthesis" as used here refers to the biosynthetic capacity of certain microorganisms to derive metabolic energy by "fixing" atmospheric nitrogen and/or carbon dioxide.

In this context we refer to these as "synthetic" microorganisms, and if they should become a predominant part of the soil microflora, then the soil would be termed a "synthetic" soil. Nitrogen-fixing microorganisms are highly diverse, ranging from "free-living" autotrophic bacteria of the genus Azotobacter to symbiotic, heterotrophic bacteria of the genus Rhizobium, and blue-green algae (now mainly classified as blue-green bacteria), all of which function aerobically. Photosynthetic microorganisms fix atmospheric carbon dioxide in a manner similar to that of green plants. They are also highly diverse, ranging from blue-green algae to green algae that perform complete photosynthesis aerobically to photosynthetic bacteria which perform incomplete photosynthesis anaerobically.

## Relationships Between Putrefaction, Fermentation, and Synthesis

The processes of putrefaction, fermentation, and synthesis proceed simultaneously according to the appropriate types and numbers of microorganisms that are present in the soil. The impact on soil quality attributes and related soil properties is determined by the dominant process. The production of organic substances by microorganisms results from the intake of positive ions, while decomposition serves to release these positive ions. Hydrogen ions play a pivotal role in these processes. A problem occurs when hydrogen ions do not recombine with oxygen to form water but are utilized to produce methane, hydrogen sulfide, ammonia, mercaptans and other highly reduced putrefactive substances most of which are toxic to plants and produce malodors. If a soil is able to absorb the excess hydrogen ions during periods of soil anaerobiosis and if synthetic microorganisms such as photosynthetic bacteria are present, they will utilize these putrefactive substances and produce useful substrates from them which helps to maintain a healthy and productive soil.

The photosynthetic bacteria, which perform incomplete photosynthesis anaerobically, are highly desirable, beneficial soil microorganisms because they are able to detoxify soils by transforming reduced, putrefactive substances such as hydrogen sulfide into useful substrates. This helps to ensure efficient utilization of organic matter and to improve soil fertility. Photosynthesis involves the photo-catalyzed splitting of water which yields molecular oxygen as a by-product. Thus, these microorganisms help to provide a vital source of oxygen to plant roots.

Reduced compounds such as methane and hydrogen sulfide are often produced when organic materials are decomposed under anaerobic conditions. These compounds are toxic and can greatly suppress the activities of nitrogen-fixing microorganisms. However, if synthetic microorganisms, such as photosynthetic bacteria that utilize reduced substances, are present in the soil, oxygen deficiencies are not likely to occur. Thus, nitrogen-fixing microorganisms, coexisting in the soil with photosynthetic bacteria, can function effectively in fixing atmospheric nitrogen even under anaerobic conditions.

Photosynthetic bacteria not only perform photosynthesis but can also fix-nitrogen. Moreover, it has been shown that, when they coexist, in soil with species of Azotobacter, their ability to fix nitrogen is enhanced. This then is an example of a synthetic soil. It also suggests that by recognizing the role, function, and mutual compatibility of these two bacteria and utilizing them effectively to their full potential, soils can be induced to a greater synthetic capacity.

Perhaps the most effective synthetic soil system results from the enhancement of zymogenic and synthetic microorganisms; this allows fermentation to become dominant over putrefaction and useful synthetic processes to proceed.

## Classification of Soils Based on the Functions of Microorganisms

As discussed earlier, soils can be characterized according to their indigenous microflora which perform putrefactive, fermentative, synthetic and zymogenic reactions and processes. In most soils, these three functions are going on simultaneously with the rate and extent of each determined by the types and numbers of associated microorganisms that are actively involved at any one time.

A simple diagram showing a classification of soils based on the activities and functions of their predominant microorganisms is presente.

Disease-Inducing Soils. In this type of soil, plant pathogenic microorganisms such as Fusarium fungi can comprise 5 to 20 percent of the total microflora if fresh organic matter with a high nitrogen content is applied to such a soil, incompletely oxidized products can arise that are malodorous and toxic to growing plants. Such soils tend to cause frequent infestations of disease organisms, and harmful insects. Thus, the application of fresh organic matter to these soils is often harmful to crops. Probably more than 90 percent of the agricultural land devoted to crop production worldwide can be classified as having disease-inducing soil. Such soils generally have poor physical properties, and large amounts of energy are lost as "greenhouse" gases, particularly in the case of rice fields. Plant nutrients are also subject to immobilization into unavailable forms.

Disease-Suppressive Soils. The microflora of disease-suppressive soils is usually dominated by antagonistic microorganisms that produce copious amounts of antibiotics. These include fungi of the genera Penicillium, Trichoderma, and Aspergillus, and actinomycetes of the genus Streptomyces. The antibiotics they produce can have biostatic and biocidal effects on soil-borne plant pathogens, including Fusarium which would have an incidence in these soils of less than 5 percent. Crops planted in these soils are rarely affected by diseases or insect pests. Even if fresh organic matter with a high nitrogen content is applied, the production of putrescent substances is very low and the soil has a pleasant earthy odor after the organic matter is decomposed. These soils generally have excellent physical properties; for example, they readily, form water-stable aggregates and they are well-aerated, and have a high permeability to both air and water. Crop yields in the disease-suppressive soils are often slightly lower than those in synthetic soils. Highly acceptable crop yields are obtained whenever a soil has a predominance of both disease-suppressive and synthetic microorganisms.

Zymogenic Soils. These soils are dominated by a microflora that can perform useful kinds of fermentations, i.e., the breakdown of complex organic molecules into simple organic substances and inorganic materials. The organisms can be either obligate or facultative anaerobes. Such fermentation-producing microorganisms often comprise the microflora of various organic materials, i.e., crop residues, animal manures, green manures and municipal wastes including composts. After these amendments are applied to the soil, their number: and fermentative activities can increase dramatically and overwhelm the indigenous soil microflora for an indefinite period. While these microorganisms remain predominant, the soil can be classified as a zymogenic soil which is generally characterized by a) pleasant, fermentative odours especially after tillage, b) favorable soil physical properties (e.g., Increased aggregate stability, permeability, aeration and decreased resistance to tillage c) large amounts of inorganic nutrients, amino acids, carbohydrates, vitamins and other bioactive substances which can directly or indirectly enhance the growth, yield and quality of crops, d) low occupancy of Fusarium fungi which is usually less than 5 percent, and e) low production of greenhouse gases (e.g., methane, ammonia, and carbon dioxide) from croplands, even where flooded rice is grown.

Synthetic Soils. These soils contain significant populations of microorganisms which are able to fix atmospheric nitrogen and carbon dioxide into complex molecules such as amino acids, proteins and carbohydrates. Such microorganisms include photosynthetic bacteria which perform incomplete photosynthesis anaerobically, certain Phycomycetes (fungi that resemble algae), and both green algae and blue—green algae which function aerobically. All of these are photosynthetic organisms that fix atmospheric nitrogen. If the water content of these soils is stable, their fertility can be largely maintained by regular additions of only small amounts of organic materials. These soils have a low Fusarium occupancy and they are often of the disease-suppressive type. The production of gases from fields where synthetic soils are present is minimal, even for flooded rice.

This is a somewhat simplistic classification of soils based on the functions of their predominant types of microorganisms, and whether they are potentially beneficial or harmful to the growth and yield of crops. While these different types of soils are described here in a rather idealized manner, the fact is that in nature they are not always clearly defined because they often tend to have some of the same characteristics. Nevertheless, research has shown that a disease-inducing soil can be transformed into disease-suppressing, zymogenic and synthetic soils by inoculating the problem soil with mixed cultures of effective microorganisms. Thus it is somewhat obvious that the most desirable agricultural soil for optimum growth, production, protection, and quality of crops would be the composite soil, i.e., a soil that is highly zymogenic and synthetic, and has an established disease-suppressive capacity. This then is the principle reason for seeking ways and means of controlling the microflora of agricultural soils.

## SUMMARY AND CONCLUSIONS

Controlling the soil microflora to enhance the predominance of beneficial and effective microorganisms can help to improve and maintain the soil chemical and physical properties. The proper and regular addition of organic amendments are often an important part of any strategy to exercise such control.

Previous efforts to significantly change the indigenous microflora of a soil by introducing single cultures of extrinsic microorganisms have largely been unsuccessful. Even when a beneficial microorganism is isolated from a soil, cultured in the laboratory, and reinoculated into the same soil at a very high population, it is immediately subject to competitive and antagonistic effects from the indigenous soil microflora and its numbers soon decline. Thus, the probability of shifting the "microbiological equilibrium" of a soil and controlling it to favour the growth, yield and health of crops is much greater if mixed cultures of beneficial and effective microorganisms are introduced that are physiologically and ecologically compatible with one another. When these mixed cultures become established their individual beneficial effects are often magnified in a synergistic manner.

Actually, a disease-suppressive microflora can be developed rather easily by selecting and culturing certain types of gram-positive bacteria that produce antibiotics and have a wide range of specific functions and capabilities; these organisms include facultative anaerobes, obligate aerobes, acidophilic and alkalophilic microbes. These microorganisms can be grown to high populations in a medium consisting of rice bran, oil cake and fish meal and then applied to soil along with well-cured compost that also has a large stable population of beneficial microorganisms, especially facultative anaerobic bacteria. A soil can be readily transformed into a zymogenic/synthetic soil with disease-suppressive potential if mixed cultures of effective microorganisms with the ability to transmit these properties are applied to that soil.

The desired effects from applying cultured beneficial and effective microorganisms to soils can be somewhat variable, at least initially. In some soils, a single application (i.e., inoculation) may be enough to produce the expected results, while for other soils even repeated applications may appear to be ineffective. The reason for this is that in some soils it takes longer for the introduced microorganisms to adapt to a new set of ecological and environmental conditions and to become well-established as a stable, effective and predominant part of the indigenous soil microflora. The important consideration here is the careful selection of a mixed culture of compatible, effective microorganisms properly cultured and provided with acceptable organic substrates. Assuming that repeated applications are made at regular intervals during the first cropping season, there is a very high probability that the desired results will be achieved.

There are no meaningful or reliable tests for monitoring the establishment of mixed cultures of beneficial and effective microorganisms after application to a soil. The desired effects appear only after they are established and become dominant, and remain stable and active in the soil. The inoculum densities of the mixed cultures and the frequency of application serve only as guidelines to enhance the probability of early establishment. Repeated applications, especially during the first cropping season, can markedly facilitate early establishment of the introduced effective microorganisms.

Once the "new" microflora is established and stabilized, the desired effects will continue indefinitely and no further applications are necessary unless organic amendments cease to be applied, or the soil is subjected to severe drought or flooding.

Finally, it is far more likely that the microflora of a soil can be controlled through the application of mixed cultures of selected beneficial and effective microorganisms than by the use of single or pure cultures. If the microorganisms comprising the mixed culture can coexist and are physiologically compatible and mutually complementary, and if the initial inoculum density is sufficiently high, there is a high probability that these microorganisms will become established in the soil and will be effective as an associative group, whereby such positive interactions would continue. If so, then it is also highly, probable that they will exercise considerable control over the indigenous soil microflora which, in due course, would likely be transformed into or replaced by a "new" soil microflora.

# Bibliography


Allen, P.: *Food For the Future: Conditions and Contradictions of Sustainability*, New York: John Wiley and Sons, 1993.

Aukland, L.: *The Effect of Selective Logging on Amphibian Diversity of Budongo Forest*, Uganda. B. A., Oxford University, 1997.

Bahati, J.: *Impact of Arboricidal Treatments on the Natural Regeneration of Species Composition in Budongo Forest*, Uganda. M. Sc., Makerere University, 1997.

Barlow K: *Ecology of Food and Nutrition*, Oceania, New Guinea, 1984.

Barrows, M.: *A Survey of the Intestinal Parasites of the Primates in Budongo Forest*, Uganda. Glasgow University, 1996.

Bauer, Steffan: *Handbook of Globalization and the Environment*, CRC Press, Delhi, 2007.

Bethell, E.: *Vigilance in Foraging Chimpanzees*. University College, London University, 1998.

Biswas, A., and Cline, S.: *Global Change: Impacts on Water and Food Security*, Springer, Heidelberg, 2010.

Burdak, L.R.: *Recent Advances in Desert Afforestation*, F.R.I., Dehra dun, 1982.

Burroughs, W.J.: *The Climate Revealed*, Cambridge University Press, Cambridge, 1999.

Carley, M., P. Jenkins : *Urban Development and Civil Society, the Role of Communities in Sustainable Cities*, Sterling, VA, Earthscan Publications, 2001.

Casado, Matt A: *Food and Beverage Service Manual*. New York: Wiley, 1994.

Chouhan, T.S.: *Desertification in the World and Its Control*, Scientific Publishers, Delhi, 1992.

Daubenmire, F.: *Plants and Environment*, New York, Wiley, 1947.

Geist, Helmut: *The Causes and Progression of Desertification*, Ashgate Publishing, Delhi, 2005.

Goodman, D.C.: *From Farming to Biotechnology: A Theory of Agro-industrial Development*, Oxford, Blackwell, 1987.

Jasra, O.P. : *Environmental Agriculture*, Sarup & Sons, Delhi, 2002.

Kataria, T N : *Plant and Crop Physiology*, Pearl Books, Delhi, 2008.

Katz, P., V. J. Scully : *The New Urbanism, Toward an Architecture of Community*, New York, McGraw-Hill, 1994.

Kumar, Arvind : *Environmental Pollution and Agriculture*, APH, Delhi, 2002.

Lakshmi, Narasaiah, M. : *Financing of Agriculture by Regional Rural Banks*, Sonali Pub, Delhi, 2008.

Lal, R., 1998. *Soil Quality and Agricultural Sustainability*. Chelsea, Mich.: Ann Arbor Press.

Magalhaes, A. R. *Impacts of Climatic Variations and Sustainable Development in Semi-arid Regions*. Proceedings of International Conference. ICID. Fortaleza, Brazil, 1992.

Mal, Pratap : *Infrastructural Development for Agriculture and Rural Development*, Mohit, Delhi, 2001.

Mellor, J. W.: *The New Economics of Growth*, Ithaca, Cornell University Press, 1976.

Mohanty, A.K. : *Entrepreneurship in Agriculture : Scopes and Opportunities*, Agrotech Pub, Delhi, 2011.

Murty, S. : *Faces and Phases of Agriculture and Industry in India*, RBSA Pub, Delhi, 2004.

Murugesan, R. : *Energy Use Efficiency in Dryland Agriculture*, Kalpaz, Delhi, 2010.

Nambiar P.T.N. : *Agricultural Statistics Techniques and Procedures*, Agrobios, Delhi, 2002.

Oliver, John E.: *Encyclopedia of World Climatology*, Springer, Delhi, 2005.

Pareek O. P.: *Advances in Horticulture: Fruit Crops,* New Delhi, Malhotra Publishing House, 1993.

Parry M.L.: *Climatic Change, Agriculture and Settlements*, Dawson Folkestone UK, 1978.

Parry M.L.: *Climatic Change, Agriculture and Settlements*, Dawson Folkestone UK, 1978.

Paul, Sharma, Vijay : *Glimpses of Indian Agriculture : Macro and Micro Aspects*, Academic Foundation, Delhi, 2008.

Perrings, C. : *Sustainable Development and Poverty Alleviation in Sub-Saharan Africa*, Botswana, New York, St. Martin's Press, 1996.

Pollock NJ: *The Concept of Food in Pacific Society: A Fijian Example*, Oceania, Fiji, 1985.

Purohit, S.S. : *Organic Farming : Biocontrol and Biopesticide Technology*, Agrobios, Delhi, 2008.

Reynolds, F.: *Chimpanzees of the Budongo Forest*, New York, Rinehart and Winston, 1965.

Shiel, D.: *The Ecology of Long-term Change in a Ugandan Rainforest*. D. Phil., Oxford University, 1997.

Simonds, J. O. : *Garden Cities 21, Creating a Livable Urban Environment*, New York, McGraw-Hill, 1994.

Sobsey M (2002) *Managing food in the home: Accelerated health gains from improved water supply.* Geneva, World Health Organization.

Tucker, Mary Evelyn, and John Grim: *Worldviews and Ecology: Religion, Philosophy, and the Environment.* Orbis Books, Maryknoll, N.Y. 1994.

Uhlir, P. F. : *Scientific Data for Decision Making Toward Sustainable Development,* Washington, D.C., National Academies Press, 2003.

Walter, B., L. Arkin : *Sustainable Cities, Concepts and Strategies for Eco-city Development*, Los Angeles, CA, EHM Eco-Home Media, 1992.

Webb, R. : *Floods, Droughts and Climate Change*, University of Arizona Press, NY, 2001.

# Index